Geographisches Institut
der Universität Kiel
ausgesonderte Dublette

ARID AND SEMI-ARID ENVIRONMENTS

A. Yair & S. Berkowicz (Editors):

ARID AND SEMI-ARID ENVIRONMENTS

Geomorphological and Pedological Aspects

Selected papers of the Workshop on Erosion, Transport & Deposition Processes, IGU, Commission Measurement, Theory & Application in Geomorphology, Jerusalem, Sede Boqer, Elat, March 29–April 4, 1987

CATENA SUPPLEMENT 14

CATENA – A Cooperating Journal of the International Society of Soil Science

Cover photo by Pantomap, Jerusalem:
Aerial view of excavated water-collecting hillslope minicatchments near Sede Boqer, in the Negev desert of Israel. Runoff from the rocky upper slopes accumulates in the minicatchments and can support trees without the need for irrigation.

CIP-Titelaufnahme der Deutschen Bibliothek

Arid and semiarid environments: geomorphological and pedological aspects / Aaron Yair & Simon Berkowicz (ed.). -
Cremlingen-Destedt: Catena, 1989
(Catena: Supplement; 14)
ISBN 3-923381-17-4
NE: Yaˆir, Aharon [Hrsg.]; Catena / Supplement

©Copyright 1989 by CATENA VERLAG, D-3302 CREMLINGEN-Destedt, W.GERMANY

All rights are reserved. No part of this publication may be reproduced, stored in a retrieval system or transmitted in any form or by any means, electronic, mechanical, photocopying, recording or otherwise, without prior permission of the publisher.

This publication has been registered with the Copyright Clearance Center, Inc. Consent is given for copying of articles for personal or internal use, for the specific clients. This consent is given on the condition that the copier pay through the Center the per-copy fee for copying beyond that permitted by Sections 107 or 108 of the U.S. Copyright Law. The per-copy fee is stated in the code-line at the bottom of the first page of each article. The appropriate fee, together with a copy of the first page of the article, should be forwarded to the Copyright Clearance Center, Inc., 27 Congress Street, Salem, MA 01970, U.S.A. If no code-line appears, broad consent to copy has not been given and permission to copy must be obtained directly from the publisher. This consent does not extend to other kinds of copying, such as for general distribution, resale, advertising and promotion purposes, or for creating new collective works. Special written permission must be obtained from the publisher for such copying.

Submission of an article for publication implies the transfer of the copyright from the author(s) to the publisher.

ISSN 0722-0723 / ISBN 3-923381-17-4

CONTENTS

Preface

G.F. Dardis
Quaternary Erosion and Sedimentation in Badland Areas of Southern Africa   1

A.C. Imeson & J.M. Verstraten
The Microaggregation and Erodibility of Some Semi-Arid and Mediterranean Soils   11

R. Jahn, D. Pfannschmidt & K. Stahr
Soils from Limestone and Dolomite in the Central Algarve (Portugal), Their Qualities in Respect to Groundwater Recharge, Runoff, Erodibility and Present Erosion   25

R. Kadmon, A. Yair & A. Danin
Relationships between Soil Properties, Soil Moisture, and Vegetation Along Loess-Covered Hillslopes, Northern Negev, Israel   43

F. López-Bermúdez & M.A. Romero-Díaz
Piping Erosion and Badland Development in South-East Spain   59

M. Madeyski & K. Banasik
Applicability of the Modified Universal Soil Loss Equation in Small Carpathian Watersheds   75

J. De Ploey
A Model for Headcut Retreat in Rills and Gullies   81

B.L. Rhoads
Longitudinal Variations in the Size and Sorting of Bed Material Along Six Arid-Region Mountain Streams   87

K.-H. Schmidt
Talus and Pediment Flatirons
Erosional and Depositional Features on Dryland Cuesta Scarps   107

P. Shaw
Fluvial Systems of the Kalahari — A Review   119

K. Stahr, R. Jahn, A. Huth & J. Gauer
Influence of Eolian Sedimentation on Soil Formation in Egypt and Canary Island Deserts   127

A. Yair & S.M. Berkowicz
Climatic and Non-Climatic Controls of Aridity: The Case of the Nothern Negev of Israel   145

Preface

This special CATENA volume is an outcome of the Workshop on Erosion, Transport and Deposition Processes in Semi-Arid and Arid Areas. The Workshop, held in Israel in the Spring of 1987 was organized as a joint venture between the International Commission on Continental Erosion (IRHS) and the Commission on Measurement, Theory and Application in Geomorphology (IGU). This cooperation reflected the need for scientific interchange between hydrologists and geomorphologists, both groups having a common interest in the study of runoff and erosion processes. The meeting improved avenues of communication and enhanced cooperation in the boundary area shared by hydrology and geomorphology.

The present volume contains selected papers dealing with geomorphic and pedologic aspects of semi-arid and arid zones. They cover a wide range of physiographic conditions, extending over four continents (North America, Europe, Asia and Africa). As might be expected, the topics covered are highly diversified. Some are theoretical or are of an applied nature while other papers deal with hillslopes or channels. Although the temporal and spatial scales differ from one paper to another, all exhibit a strong interest towards the understanding of processes shaping the land.

Aaron Yair
Simon M. Berkowicz

Jerusalem, January 1989

QUATERNARY EROSION AND SEDIMENTATION IN BADLAND AREAS OF SOUTHERN AFRICA

G.F. **Dardis**, Umtata

Summary

Badland areas in eastern southern Africa are presently dominated by subsurface piping, subsurface cavitation, tunnel erosion and gully development, down to and into bedrock. Sedimentological and stratigraphic evidence from Quaternary colluvial sequences in these areas suggests that these forms are of recent age, formed in the past 2000 years. Similar forms have not been identified in colluvial sequences up to 25,000 years in age. Colluvial sequences older than late Holocene show little evidence of palaeopiping, dissection or related evidence of badland development. The older sequences are dominated by deposits formed by sheetwash and sediment gravity flow, reflecting micropedimentation processes rather than accelerated erosion.

1 Introduction

The development of badland topography is most commonly associated with incipient instability within a fluvial system in areas which share specific lithologic, climatic and land use characteristics (SCHEIDEGGER et al. 1968, BRYAN & YAIR 1982). Many different factors can act to trigger a complex sequence of events that will result in soil erosion and gully development. Even though badlands are ubiquitous in southern Africa, little is known of when soil erosion commenced, or whether changes in the nature and pattern of erosion have occurred over time. Long-term climate-geomorphic responses (cf. WELLS & GUTIERREZ 1982) need to be examined in southern Africa before factors contributing to soil erosion (i.e. soil texture, surface and subsurface hydrology, man etc.) can be properly evaluated. The aim of this paper is to determine whether processes operating in the past in badland areas of southern Africa were different from the present, by comparing contemporary badland processes operating in southern Africa with evidence found in Quaternary sedimentary successions in badland areas.

2 Contemporary badland processes in southern Africa

Nine major contemporary soil erosion forms have been identified in southern Africa (BECKEDAHL et al. 1988, DARDIS et al. 1988). Of these, four are

ISSN 0722-0723
ISBN 3-923381-17-4
©1989 by CATENA VERLAG,
D–3302 Cremlingen-Destedt, W. Germany
3-923381-17-4/89/5011851/US$ 2.00 + 0.25

Photo 1: *Type 1 soil erosion form in bushveld terrain, Swaziland.*

Photo 2: *Type 6/7 soil erosion form, Inxu Drift, Transkei.*

Photo 3: *Type 8 soil erosion form, middle highveld region, Swaziland.*

probably the most widespread:

1. Type 1 erosion forms (photo 1), affecting parts of the Karoo (ROWNTREE 1988), and bushveld and thornveld terrain (notably in Ciskei and Swaziland). These forms result from water-activated unconfined erosion (e.g. sheetwash, sheetflooding, rainsplash erosion) and are generally associated with areas where vegetation is absent (or removed), or with surfaces which inhibit infiltration (ROWNTREE 1988, DARDIS et al. 1988).

2. Type 2 forms, resulting from aeolian activity, affecting sand-covered areas in the interior, and sand-dominant coastlines in southern Africa (TINLEY 1984, THOMAS 1988, ROWNTREE 1988, DARDIS & GRINDLEY 1988).

3. Type 6/7 soil erosion forms (photo 2), mainly affecting colluvium-mantled terrain throughout eastern southern Africa (PRICE-WILLIAMS et al. 1982, WATSON et al. 1984, HOOKER 1984, BECKEDAHL 1977, BECKEDAHL & DARDIS 1988, DARDIS & BECKEDAHL 1988a). These forms are associated with heterogeneous materials and form primarily as a result of collapse of subsurface soil pipes and cavities, and by backwearing processes (DARDIS et al. 1988).

4. Type 8 soil erosion forms (photo 3) occur in the coastal fringe of eastern southern Africa (HOOKER 1984), in the Drakensberg, and in areas of deep weathering, particularly in the eastern Transvaal (LAGEAT & ROBB 1984). They form mainly

in homogeneous materials, such as weathered bedrock, coastal sands, and soft bedrock (DARDIS & BECKEDAHL 1988b, 1988c). They may also develop in massive and poorly stratified colluvium. They are associated primarily with surface backwearing processes (DARDIS et al. 1988).

3 Quaternary erosion and sedimentation

In most of the erosion forms (described above), it is generally difficult to establish when erosion commenced, or to assess changes (if any) in the nature and pattern of erosion with time, except where well-defined stratigraphic marker horizons have been dissected in the eroded matrial.

Type 6/7 erosion forms, which generally dissect colluvial deposits, provide one of the best means of assessing possible changes with time. They occur over a wide area, have a wide altitudinal range and, in southern Africa, commonly contain relatively long time-stratigraphic sequences (in the range 25–40 ka) (cf. WATSON et al. 1984). Depositional sequences in type 6/7 erosion forms have been examined at over 180 sites in eastern southern Africa. The majority of these sites occur either in hillside hollows or valley bottoms. These are sites of topographically-induced convergence, and form natural depositories for colluvial debris. Changes in storage or discharge of materials at these sites has been assessed principally on

1. the nature of the lithofacies associations (e.g. colluvial-periglacial, colluvial-alluvial etc.) present in colluvial sequences,

2. on characteristic bed contacts between stratigraphic units, and

3. on age determinations of the sequences.

The type of processes (e.g. sheetwash, piping, seepage erosion, sediment flowage, gullying etc.) operating during periods of degradation has been assessed principally on the nature of bed contacts within the colluvial sequences. Erosion rates have been estimated only where multiple dates have been obtained from particular sites.

A schematic model (fig.1) summarizes the bed contact characteristics found in colluvial sequences associated with type 6/7 soil erosion forms. Present day piping and rill and gully erosion is apparent in these sequences. In most cases the incision occurs down to and into underlying bedrock. Bed contacts are normally planar or gradational. There is no evidence of palaeopiping or other erosional topography on bed contacts in the underlying, older colluvium units, which presumably record an hiatus in sedimentation (i.e. an erosional episode). The sites commonly show several phases of erosion and sedimentation, and similar bed contact relationships are apparent, regardless of variations in sediment type. Even where deposition has occurred on high-angle slopes, bed contacts between stratigraphic units remain planar or gradational. Cross-cutting relationships between beds is minimal. No evidence of palaeopiping or cut-and-fill structures is apparent within either the colluvium or associated deposits. Bed contacts show no evidence of internal faulting or disturbance.

Dating of these deposits at a number of sites suggests that several phases of colluviation and subsequent erosion

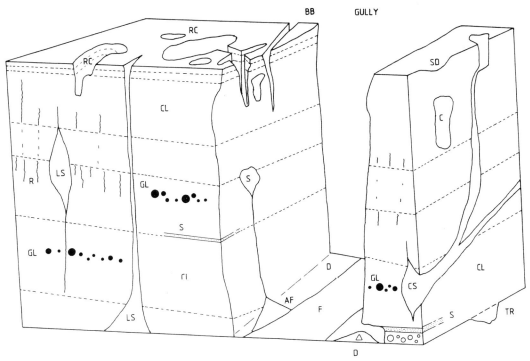

Fig. 1: *A schematic model of present day gully erosion in Transkei.*
CL: Colluvium, S: Soil pipe, LS: Soil pipe associated with linear cracking, CS: Composite soil pipe network, BB: block breakage associated with surface cracking and gullying, TR: Tunnel erosion in bedrock, associated with piping, C: Cavity, RC: Roof collapse features, SD: Surface depression, R: Rills, P: Palaeosol, GL: Boulder lag, S: Sand layer, D: Debris, with mud balls, AF: Alluvial cone, F: Fluvial erosion and deposition.

have occurred in the past 25,000 years (PRICE-WILLIAMS et al. 1982, WATSON et al. 1984, DARDIS in prep.). Evidence from bed contacts suggests that no major dissection has occurred in this time, though there is evidence for a number of erosional phases. These erosion phases appear to have been dominated by sheetwash, sediment flowage and, in some instances, fluvial processes. The combination of inferred processes is similar to those found in modern day pediments in arid and semi-arid environments (DENNY 1967, WILLIAMS 1974, SWAN 1974) and miniature pediments formed in badland areas (SMITH 1958, SCHUMM 1956, 1962). This contrasts with the present day, where much of the erosion is associated with gully development.

Radiocarbon dating of organic-rich paleosols in Transkei (DARDIS, in prep.) suggests that the present phase of degradation commenced 2000 years ago. Radiocarbon dating of wood fragments in colluvial sequences in Swaziland (DARDIS et al., in prep.) also suggests that the incision phase may be much younger, dating to the last millenium, and shows a rapid increase in erosion rates in the last 250 years. The type of erosion occurring on the present

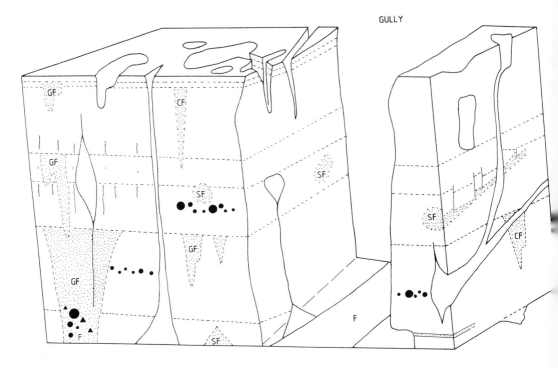

Fig. 2: *A hypothetical model fo gully erosion, assuming Quaternary erosion cycles similar to present day denudation processes, with post-depositional infilling of erosion scars.*
GF: Gully infilling, SF: Soil pipe infilling, CF: Crevasse filling, F: Fluvial erosion and deposition, Triangles: Debris infilling

land surface (fig.1) is not evident during either erosion phases in the past 25,000 years. If it had occurred, evidence of palaeopiping of infilling (fig.2) or, in the absence of sediment infilling erosional topography, key stratigraphic evidence from bed contacts (fig.3) should be forthcoming in Quaternary sedimentary successions. Few of the features postulated in figs.2 and 3 have yet been reported from Quaternary successions in southern Africa, to suggest that older phases of incision have occurred. Similar observations have been reported in the American southwest, where there is little evidence of arroyo-cutting prior to 1880 (ANTEVS 1952, COOKE & REEVES 1976).

Older (i.e. greater than 30,000 years) colluvial sequences occurring as erosional remnants have been identified in southern Africa (cf. PRICE-WILLIAMS et al. 1982, WATSON et al. 1984). These remnants are generally too fragmentary to allow detailed evaluation of their initial thicknesses or establish the nature of post-depositional processes responsible for their removal. As much of the Quaternary colluvial sedimentary record appears to be younger than 25,000 years, the paucity of older deposits poses the problem of whether erosion, rather

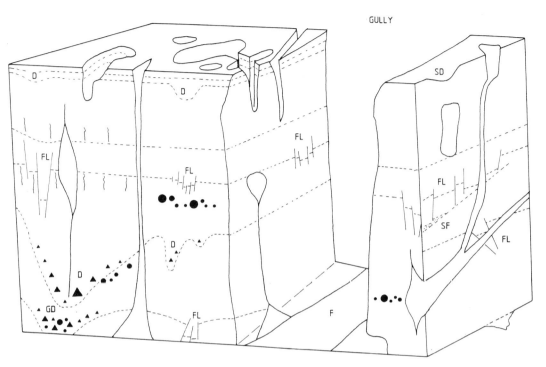

Fig. 3: *A hypothetical model of gully erosion, assuming a series of Quaternary erosion cycles similar to present day denudation processes, with no post-depositional infilling of erosion scars.*
D: Debris infilling, FL: Faulting, SD: Surface depression, F: Fluvial erosion and deposition, SF: Soil pipe infilling, Triangles: Boulder/debris lag, GD: Relict fluvial gravel facies interbedded with debris infillings

than deposition, was the norm during the Quaternary in eastern southern Africa.

4 Conclusions

The fundamental problem of establishing causes of soil erosion and gully development has not been addressed in this paper. It is clear, however, that the type of erosion that is occurring on the present land surface does not appear to be well-documented in late Quaternary colluvial sequences, and is essentially a recent, widespread geomorphic event. While it has not been possible to accurately date the onset of incision in other soil erosion forms, it seems highly likely that these are also recent, and perhaps occurring at a much faster rate (DARDIS et al. 1988). Extensive type 8 soil erosion forms are developed in deeply weathered granitoid rocks (up to 50 m depth in places) in the middle highveld region of Swaziland and in the eastern Transvaal (DARDIS & BECKEDAHL 1988b, 1988c). This demonstrates recent instability on slopes which previously have been stable for a sufficient priod of time to allow deep weathering to take place.

References

ANTEVS, E. (1952): Arroyo-cutting and filling. Fournal of Geology, 60, 375–384.

BECKEDAHL, H.R. (1977): Subsurface erosion near the Oliviershoek Pass, Drakensberg. South African Geographical Journal, 59, 130–138.

BECKEDAHL, H.R. & DARDIS, G.F. (1988): The role of artificial drainage in the development of soil pipes and gullies: Some examples from Transkei, southern Africa. In: Geomorphological Studies in Southern Africa (Eds. G.F. DARDIS & B.P. MOON), A.A. Balkema, Rotterdam, 229–245.

BECKEDAHL, H.R., BOWYER-BOWER, T.A.S., DARDIS, G.F. & HANVEY, P.M. (1988): Geomorphic effects of soil erosion. In: The Geomorphology of Southern Africa. (Eds. B.P. MOON & G.F. DARDIS), Southern Book Co., Johannesburg, 249–276.

BRYAN, R.B. & YAIR, A. (1982): Perspectives on studies of badland geomorphology. In: Badland Geomorphology and Piping. (Eds. R.B. BRYAN & A. YAIR), Geo Books, Norwich, 1–12.

COOKE, R.U. & REEVES, R.W. (1976): Arroyos and Environmental Change in the American South West. Clarendon Press, Oxford, 213 pp.

DARDIS, G.F. (in prep.): Age of Quaternary colluvium in eastern southern Africa.

DARDIS, G.F. & BECKEDAHL, H.R. (1988a): Drainage evolution in an ephemeral soil pipe-gully system, Transkei, southern Africa. In: Geomorphological Studies in Southern Africa. (Eds. G.F. DARDIS & B.P. MOON), A.A. Balkema, Rotterdam, 247–265.

DARDIS, G.F. & BECKEDAHL, H.R. (1988b): Gully formation in Archaean rocks at Saddleback Pass, Barberton Mountain Land, South Africa. In: Geomorphological Studies in Southern Africa. (Eds. G.F. DARDIS & B.P. MOON), A.A. Balkema, Rotterdam, 285–297.

DARDIS, G.F. & BECKEDAHL, H.R. (1988c): Granite landforms in the middle highveld region of Swaziland. In: Southern African Landscapes: A Geomorphological Field Guide. (Eds. H.R. Beckedahl, B.P. MOON & G.F. DARDIS), Department of Geography, University of Transkei, 209–215.

DARDIS, G.F. & GRINDLEY, J.R. (1988): Coastal geomorphology. In: The Geomorphology of Southern Africa. (Ed. B.P. MOON & G.F. DARDIS), Southern Book Co., Johannesburg, 166–195.

DARDIS, G.F., BECKEDAHL, H.R., BOWYER-BOWER, T.A.S. & HANVEY, P.M. (1988): Soil erosion forms in southern Africa. In: Geomorphological Studies in Southern Africa. (Eds. G.F. DARDIS & B.P. MOON), A.A. Balkema, Rotterdam, 187–213.

DARDIS, G.F., BECKEDAHL, H.R. & PRINS, F. (in prep.): Late Holocene erosion and colluvium depisition in Swaziland.

DENNY, C.S. (1967): Fans and pediments. American Journal of Science, 265, 81–105.

HOOKER, R.M. (1984): Gully (donga) erosion in Swaziland — A distributional analysis. Unpublished manuscript.

LAGEAT, Y. & ROBB, L.J. (1984): The relationships between structural landforms, erosion surfaces and the geology of the Archaean granite basement in the Barberton region, eastern Transvaal. Transactions, Geological Society of South Africa, 87, 141–159.

PRICE-WILLIAMS, D., WATSON, A. & GOUDIE, A.S. (1982): Quaternary colluvial stratigraphy, archaeological sequences and palaeoenvironments in Swaziland, southern Africa. Geographical Journal, 148, 50–67.

ROWNTREE, K.M. (1988): Equilibrium concepts, vegetation change and soil erosion in semi-arid areas: Some considerations for the Karoo. In: Geomorphological Studies in Southern Africa. (Eds. G.F. DARDIS & B.P. MOON), A.A. Balkema, Rotterdam, 175–185.

SCHEIDEGGER, A.E., SCHUMM, S.A. & FAIRBRIDGE, R.W. (1968): Badlands. In: The Encyclopedia of Geomorphology. (Ed. R.W. FAIRBRIDGE), Reinhold, New York, 43–48.

SCHUMM, S.A. (1956): The role of creep and rainwash in the retreat of badland slopes. American Journal of Science, 254, 693–706.

SCHUMM, S.A. (1962): Erosion on miniature pediments in Badlands National Monument, South Dakota. Geological Society of America Bulletin, 73, 719–724.

SMITH, K.G. (1958): Erosion processes and landforms in Badlands National Monument. Geological Society of America Bulletin, 69, 975–1008.

SWAN, S.B.St.C. (1974): Piedmont slope studies in a humid tropical region, Johor, southern Malaya. Zeitschrift für Geomorphologie, Suppl. Bd. 10, 30–39.

THOMAS, D.S.G. (1988): The geomorphological role of vegetation in dune systems of the Kalahari. In: Geomorphological Studies in Southern Africa (Eds. G.F. DARDIS & B.P. MOON), A.A. Balkema, Rotterdam, 145–157.

TINLEY, K. (1984): Coastal Dunes of South Africa. South African national scientific programmes report **109**, CSIR, Pretoria.

WATSON, A., PRICE-WILLIAMS, D. & GOUDIE, A.S. (1984): The palaeoenvironmental interpretation of colluvial sediments and palaeosols of the late Pleistocene hypothermal in southern Africa. Palaeogeography, Palaeoclimatology, Palaeoecology, **45**, 225–249.

WELLS, S.G. & GUTIERREZ, A.A. (1982): Quaternary evolution of badlands in the southern Colorado Plateau, U.S.A. In: Badland Geomorphology and Piping. (Eds. R.B. BRYAN & A. YAIR), Geo Books, Norwich, 239–258.

WILLIAMS, G.E. (1974): Piedmont sedimentation and late Quaternary chronology in the Biskra region of the northern Sahara. Zeitschrift für Geomorphologie, Suppl. Bd. **10**, 40–63.

Address of author:
George F. Dardis
Department of Geography
University of Transkei
Private Bag X5092
Umtata, Transkei, southern Africa
Present Address:
8 Drum Road
Cookstown, Northern Ireland

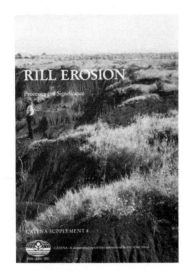

R. B. Bryan (Editor)

RILL EROSION

Processes and Significance

CATENA SUPPLEMENT 8
192 pages / hardcover / price DM 149,— / US $ 88.—
Special rate for subscriptions until
December 15, 1987: DM 119,— / US $ 70.40

Date of publication: July 15, 1987   ORDER NO. 499/00107

ISSN 0722-0723/ISBN 3-923381-07-7

CONTENTS

R.B. BRYAN
PROCESSES AND SIGNIFICANCE OF RILL DEVELOPMENT

G. GOVERS
SPATIAL AND TEMPORAL VARIABILITY IN RILL DEVELOPMENT PROCESSES AT THE HULDENBERG EXPERIMENTAL SITE

J. POESEN
TRANSPORT OF ROCK FRAGMENTS BY RILL FLOW—A FIELD STUDY

O. PLANCHON, E. FRITSCH & C. VALENTIN
RILL DEVELOPMENT IN A WET SAVANNAH ENVIRONMENT

R.J. LOCH & E.C. THOMAS
RESISTANCE TO RILL EROSION: OBSERVATIONS ON THE EFFICIENCY OF RILL EROSION ON A TILLED CLAY SOIL UNDER SIMULATED RAIN AND RUN-ON WATER

M.A. FULLEN & A.H. REED
RILL EROSION ON ARABLE LOAMY SANDS IN THE WEST MIDLANDS OF ENGLAND

D. TORRI, M. SFALANGA & G. CHISCI
THRESHOLD CONDITIONS FOR INCIPIENT RILLING

G. RAUWS
THE INITIATION OF RILLS ON PLANE BEDS OF NON-COHESIVE SEDIMENTS

D.C. FORD & J. LUNDBERG
A REVIEW OF DISSOLUTIONAL RILLS IN LIMESTONE AND OTHER SOLUBLE ROCKS

J. GERITS, A.C. IMESON, J.M. VERSTRATEN & R.B. BRYAN
RILL DEVELOPMENT AND BADLAND REGOLITH PROPERTIES

THE MICROAGGREGATION AND ERODIBILITY OF SOME SEMI-ARID AND MEDITERRANEAN SOILS

A.C. **Imeson** & J.M. **Verstraten**, Amsterdam

Summary

The size distributions of water-stable silt and clay fractions are examined for soils from three locations in Spain and considered with respect to soil erodibility, soil aggregation and/or erosion.

A comparison of water-stable fractions in regolith materials or sediments on badland slopes and micropediments at Dahesas Granada showed an enrichment of coarser fractions on the pediment in a chemical environment in which clay particles are flocculated. Large aggregates or fragments of crust are more stable on the micropediment than on the badland slopes.

The importance of organic matter in producing coarser more stable soil aggregates in highly calcareous soils is demonstrated for several locations in Alicante Province. Fine water stable particles released by slaking are segregated by splash action and form small crusted surfaces.

For soils under cultivation and forest on granodiorite in Gerona Province water-stable silt and clay fractions are compared to the fractions obtained when standard procedures are used to remove organic matter and carbonates. The largest differences are generally in the fractions <10 μm, indicating that these are affected most by stabilizing substances. A stability index is calculated based on the % water-dispersible silt and clay (% WDSC) which can be compared to aggregate stability measurements. Where soils are cultivated, have high amounts of water-soluble salts, or well developed textural B horizons, these have water-dispersible size fractions which deviate from the general relationships found.

1 Introduction

The way in which soils or regolith materials react to wetting by rainfall is of great importance for erosion. This reaction involves both physico-chemical and mechanical responses which may result in partial or complete slaking, swelling and dispersion. In earlier papers GERITS et al. (1987) and IMESON & VERSTRATEN (1988) described how the dynamic response of materials to wetting could be used to explain the distribution and occurence of rills on badland slopes in southeast Spain. In this paper attention is given to the size-distribution of the micro-aggregates into which moistened soils break down. This is done for erodible, highly calcareous soils from

ISSN 0722-0723
ISBN 3-923381-17-4
©1989 by CATENA VERLAG,
D-3302 Cremlingen-Destedt, W. Germany
3-923381-17-4/89/5011851/US$ 2.00 + 0.25

southeast Spain and for relatively less erodible soils developed under forest on granodiorite and granitic parent materials from northeast Spain, which become highly erodible under cultivation. The paper is mainly based on determinations of the water stable silt and clay fractions and on the micromorphology of thin sections.

There are several reasons for studying the size-distribution of water stable particles. For example, there are very many observations of eroded material being transported in an aggregated state and it is known that physical properties of various soils are related to their water stable clay content (SHANMUGANATHAN & OADES 1982, CHITTLEBOROUGH 1982). The development of surface crusts, the infiltration process and the availability of material for mechanical transport are all influenced by soil microaggregation. In addition, since soil microaggregation is sensitive to changes in soil organic matter, it is probably a good indicator of soil degradation. In the case of highly calcareous and/or gypsiferous soils, studying the water stable silt and clay fractions is a way of avoiding the difficulty of standard particle size analysis (IMESON & VERSTRATEN 1985).

The objective of this paper is to illustrate some of the relationships which have been found between the response of soil to wetting, erodibility, and the water stable silt and clay fractions, during the course of research at various locations in Spain. Three field situations have been selected as examples; the first is from the Rio Fardes badlands in the province of Granada; the second is from an area of calcareous marls near Teulada in Alicante; and the third from an area of granitic and granodioritic rocks in the Province of Gerona. At the first location the microaggregation of badland slope and micropediment surface materials is compared, at the second the microaggregation of material from forested and cultivated sites is examined and at the last site the microaggregation of relatively stable Mediterranean forest soils formed on granodiorite is compared with that developed on other materials where the soil is more erodible. Details of the climate, geology and vegetation of the three sites are described in other publications (GERITS et al. 1987, IMESON & VERSTRATEN 1985, SEVINK 1988).

2 Methods

For the highly calcareous soils, the size distribution of the water stable silt and clay fraction was determined by shaking a 25 g sample of the fine earth fraction for one hour in 0.5 l of distilled water, transferring the suspension to a sedimentation jar, making the sample up to one litre, homogenizing it and determining the size distribution of material with an equivalent diameter <50 μm by pipette analysis (IMESON & VERSTRATEN 1985). For the samples from northeast Spain not containing any carbonates in the solum, the size distribution obtained in this way could be compared with that found by standard grain size analysis with organic matter removed. For some samples thin sections were made to examine the aggregation of samples microscopically.

For all of the samples, analyses were undertaken to determine pH, free iron, organic carbon, water soluble salts in saturation extracts (K, Na, Ca, Mg, Cl, NO_3, SO_4, ortho-P). Aggregate stability was assessed in various ways using ultrasonic and water-drop test procedures

Microaggregation and Erodibility, Semi-arid Soils

Photo 1: *Badland landscape at Dehesas.*

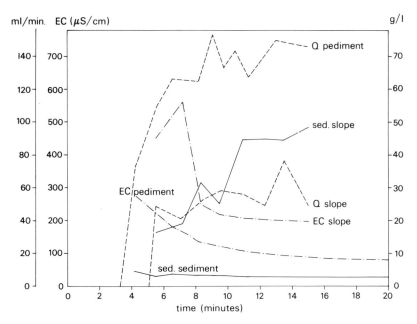

Fig. 1: *A comparison of the response of micropediment and badland slope surfaces at the Dehesa site when subjected to identical rainfall simulation runs.*

(IMESON & VIS 1984). The details of these methods and of the rainfall simulation experiments referred to, are given in GERITS et al. (1987) and IMESON & VERSTRATEN (1985).

3 Results and discussion

3.1 The aggregation of material from the Dehesas badland slopes, Granada

The surficial regolith horizons in the Rio Fardes badlands at the Dehesas field site have been described in detail by FINLAYSON et al. (1986) and by GERITS (1989). The various regolith horizons are composed mainly of silt size calcium carbonate particles (60–70%) and contain only 25 to 35% of clay and silt size silicate minerals. There is very little material coarser than 50 μm. The two most important elements of the relief are the steep badland slopes and the micropediments in front of them (photo 1).

Rainfall simulation experiments had shown that material on these two units responded very differently to simulated rainfall. The badland slope released high concentrations of sodium and suspended sediment and in general produced water with high practical sodium adsorption ratio (SAR_p) values conducive to the dispersion of clay. The pediment produced relatively more runoff but this contained much lower sediment and solute concentrations (fig.1). Samples analyzed from micropediments and badland slope surfaces had shown that the former had a higher stability (IMESON & VERSTRATEN 1985).

To examine whether the differences in material properties were general, samples were collected at four paired sites one metre above and below the badland slope-micropediment boundary. The aggregate stability and water stable silt and clay fractions were determined and thin sections made from undisturbed samples.

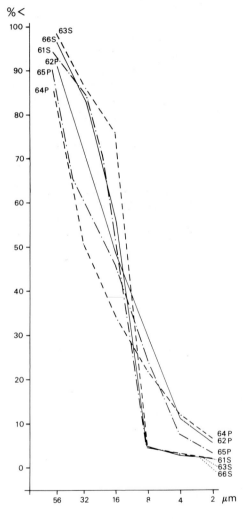

Fig. 2: *The cumulative water-stable size distribution of surface (0–2 cm) material from micropediment (P) and badland slopes (S) sites, Dehesa.*

The water stable silt and clay fractions (fig.2), show consistent differences between micropediment and badland slope sites. The fractions <8 μm and >32 μm are more important on the pediment. The important 16-8 μm fraction which forms between 50 and 72 percent of the untransported badland regolith material varied between only 12 and 22% on the micropediment. An enrichment of material >32 μm might be expected due to the selectivity of hydraulic transport. The enrichment of the finer fraction is difficult to explain unless the physico-chemical environment is taken into account. The clay present on the slope above the pediment can be seen in thin section to be present in dispersed bands. This is consistent with the chemistry of the water soluble salts of the saturated pastes (low SAR_p with respect to the EC25 value on the pediment; high SAR_p with respect to the EC25 value on the backslope).

Several thin sections from the badland surface material showed signs of transportation, presumably by flow processes, but others seemed to be composed of material weathered in situ. All of the samples from the micropediment showed depositional crusts with repeated sequences of coarser and finer aggregates.

The aggregate stability of the micropediment samples was higher in all cases than on the badland slopes. The aggregates tested from the micropediment consisted of broken fragments of depositional crusts. These did not lose material by dispersion as did the badland slope samples.

The badland slope and micropediment are separated by a physico-chemical discontinuity. Two micro-environments having contrasting response seem to have developed. Due to the structure of the micropediment surface, a low rate of infiltration exists and runoff might be expected more frequently and for longer periods than on the badland slope. The low rate of infiltration is accompanied by a low rate of sediment release. Fine silt and flocculated clay (silicate) minerals together with the coarse silt fraction (carbonates) accumulate on the pediment, while the intermediate silt fractions (carbonates) are preferentially removed.

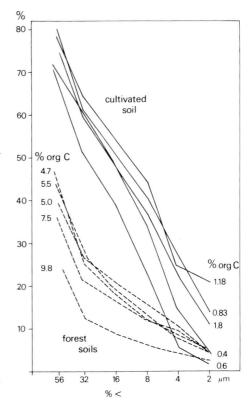

Fig. 3: *The cumulative water stable size distribution of surface (0-2 cm) material from cultivated and forest soils near Teulada, Alicante.*

sample	first rainfall event air dry sample	% <1 mm	first rainfall event sample at pF1	% <1 mm	second rainfall event	% >1 mm
31	7	100	36	50	79	40
32	102	10	—	20	136	20
36	140	10	208	20	100	20
38	241	50	171	30	226	30

Tab. 1: *Amount of material (mg/mm rain) splashed from samples during experiment 1 and the proportion of material finer than 1 mm on the surface.*

3.2 The microaggregation of highly calcareous soils near Teulada, Alicante

In a previous paper IMESON & VERSTRATEN (1985) described the contrasting size distributions of water-stable silt and clay fractions of samples from the province of Alicante. The much finer aggregation of water-stable particles from cultivated soils and the association between aggregate size and organic matter content is illustrated in fig.3. Tilled soil is composed of coarse aggregates which slake and fuse together under rainfall and eventually form crusts. Aggregates sampled from the surface soil are composed of silt sized calcium carbonate particles of various sizes and are sometimes rather porous. The small amount of organic material observed in thin section consists of coarse plant remains and this does not contribute to the stability of soil aggregates.

Two experiments were carried out to gain an insight into the crusting of these soils. In the first experiment samples from five Ap horizons were subjected, under moist or dry states, to either one or two rainfall simulation experiments. The amount of material splashed from the sample during the experiment was determined and thin sections made of the treated samples. In the second experiment the effect of a number of raindrop impacts on the structure of soil aggregates was examined.

3.2.1 Experiment one

The procedure followed was to gently place material <2.8 mm into 5 cm diameter rings originally constructed for pF determinations. Two samples were moistened to pF 1 for 24 hours and one sample left air-dry. Each sample was then subjected to simulated rainfall for ten minutes at an intensity of between 50 to 60 mm h^{-1}. The splashed material was caught and weighed. The samples were dried under infrared lamps for 2 hours and then air dried. The rainfall simulation was repeated for one of the samples after a two hours drying period. Samples 32 and 38 (tab.1) were impregnated with resin and thin sections were prepared following MÜCHER & DE PLOEY (1984).

It was thought that the air-dry sample would be more sensitive to slaking than the moistened samples and consequently would show more crusting, and also that after two rainfall events crusting would be better developed. The thin sections in all cases showed all of the samples to be composed of a range of silt-sized calcium carbonate particles. The simulated rainfall on the air-dried samples had little effect in the case of sample 32 which had a surface that remained loose and porous. In the case of sample 38, however, there was some evidence of the development of a slightly compact layer of particles having a platey structure (tab.1). The thin sections for samples 32 and 38, made after one rainfall event, showed a thin layer of splashed fine particles on a somewhat more compact layer below. Both samples were similar but more segregation of material had occurred in the case of sample 38 where finer material had accumulated in micro-depressions. Between the depressions, coarser particles were present and these seemed to have protected the underlying finer particles from compaction. Samples that had undergone two rainfall events showed slightly more compaction, but the surfaces were in both cases quite porous. Differences in samples subjected to one rainfall event were not great.

3.2.2 Experiment two

In this experiment the objective was to examine the effect of raindrop impact on aggregates 4–5 mm in size. The aggregates were subjected to ten impacts by drops, 0.1 g in weight, falling a distance of one metre. Two similar A horizon samples were selected (31 and 36) and from each sample ten aggregates were treated. The thin sections showed original aggregates to be rahter massive, slightly more so for sample 36, and to be composed of finer grains. Following the drop impacts, the material composing the aggregate appeared to increase in porosity; finer material presumably being lost. The degree of increase in porosity was variable, some aggregates hardly being affected.

The above experiments, and the analyses of the water stable silt fractions of these soils which have very low aggregate stability and which show extensive crust development (IMESON & VERSTRATEN 1985), establish that the water stable fractions are formed by calcium carbonate particles which are only weakly aggregated. During rainfalll, crusts are formed by compaction and by the segregation of finer and coarser particles. The finer particles can be protected by coarser aggregate grains, but the coarser aggregates themselves rapidly lose finer material as they become porous, lose cohesion and collapse into primary grains. It can be seen that this does not occur when iron oxides or organic material are present, provided that the latter docs not consist of coarse plant remains.

3.3 The microaggregation of forest soils on granodiorite near Gerona

Lage areas of Mediterranean forest occur to the southeast of Gerona on hilly areas of post Hercynian granite and granodiorite. For almost 40 sites having similar topographic positions, the A and B horizons of the soil were described and details of the forest vegetation recorded. Full details are to be reported in a later publication. At most sites the forest was dominatd by oak (*Q. suber* and *Q. ilex*) and/or pine (*P. halepensis* and/or *P. pinaster*) and there was an understorey-dominated tree heather (*Erica arborea*) and *Arbutus unedo*.

The initial objectives of the research were to examine the influence of atmospheric salt on soil structure and to examine the influence of forest composition on soil aggregation. In this part the microaggregation grain size distribution and aggregate stability of some of these samples will be described.

Most of the sites are located on the

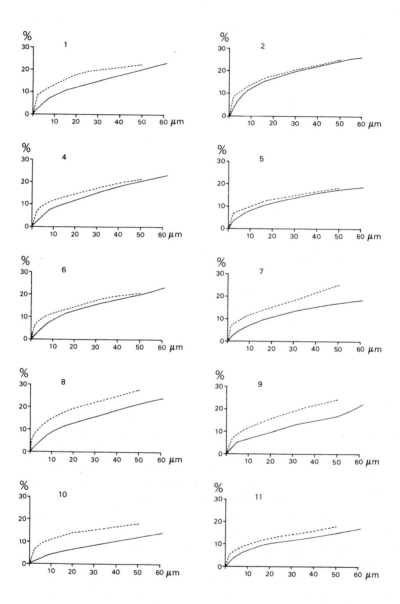

Fig. 4: *A comparison of the cumulative size distributions obtained by standard grain size analyses (stippled line) and by the procedure for determining the water stable silt and clay fractions (solid line).*
The left and right hand figures are respectively A and Bw horizons from the same profiles on granodiorite.

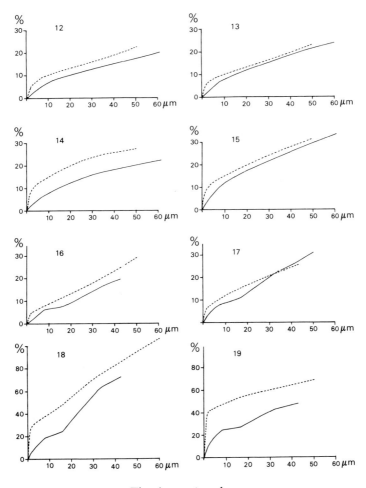

Fig. 4: *continued.*

upper parts of pediment slopes where an horizon of variable thickness lies on a shallow Bw horizon which passes into granodiorite gruss at a depth of about 30 to 40 cm. This is the case for samples 1 to 17, examples from which are included in fig.4. The other samples are from the same area but are on Early Pleistocene sediments largely derived from the granodiorite. In one case (samples 18 and 19) the A and B horizons contain, respectively 15 and 47% of illuviated clay compared to values of less than 9% for the granodiorite sites. Sample 24 to 30 (fig.5) are from a palaeosol which is frequently exposed in the region and which contains dispersible clay in the BC horizon (sample 28 and 29) where the percentage of soluble sodium with respect to the sum cations and consequently the exchangeable sodium percent is very high. Analytical data for a number of representative soils are given in tab.2.

The water stable silt and clay fractions

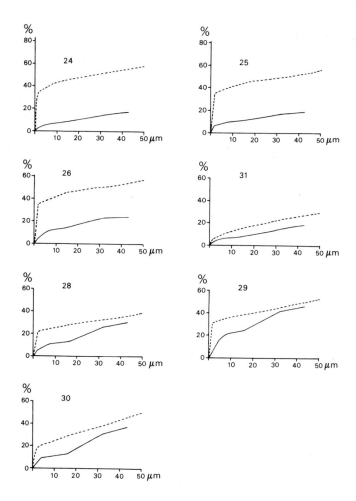

Fig. 5: *Samples from soil profile developed in Pleistocene deposits derive from granodiorite. For analyses see tab.2.*
Stippled line: cumulative size distribution obtained from standard grain size analysis.
Solid line: cumulative size distribution of water stable silt and clay fractions.

are plotted as cumulative graphs in fig.4, together with the grain size distributions obtained with pre-treatment to remove carbonates and organic matter. The left and right hand figures are from the A and Bw horizons respectively. It can be seen that the grain size distribution of the A and Bw horizons on granodiorite are in general rather similar, indicating the importance of parent material. This is obviously not the case for the samples where clay illuviation is important (for example samples 18 and 19).

The relative stability of the soil microaggregates can be seen by comparing the cumulative grain size and water stable fraction curve for each sample. The greater the distance between the two curves, the greater the degree of water stable aggregation. For example,

Sample	horizon	% org. carbon	% free iron*	Na**	$\sum$Cat**	% WDSC***
1	A	2.4	1.15	0.10	0.57	22
2	BW	0.6	1.34	0.60	0.87	27
3	C	2.4	1.46	0.42	0.65	9
4	A	3.2	1.28	0.08	0.38	23
5	Bw	0.8	1.29	0.10	0.3	18
6	A	0.8	0.96	0.07	0.26	23
7	Bw	0.9	1.07	0.07	0.18	18
8	A	2.9	0.8	0.06	0.81	24
9	B	0.8	0.76	0.04	0.17	22
10	A	1.4	0.95	0.05	0.25	14
11	B	0.9	1.07	0.05	0.16	17
12	A	2.2	0.49	0.06	0.34	21
13	C	1.7	0.51	0.06	0.23	24
14	A	4.3	1.12	0.05	0.41	22
15	B	1.3	1.35	0.04	0.15	33
24	A	4.9	1.26	0.05	1.72	17
25	B1	2.1	1.23	0.04	0.91	19
26	B2	1.4	1.33	0.04	0.81	25
28	BC	0.2	1.53	0.04	0.18	30
29	BC	0.1	1.95	0.13	0.18	47
30	Ap	0.7	1.35	0.05	1.72	37

* as Fe_2O_3
** water soluble cations in meq/100 g abs. dry soil

*** % WDSC = $\dfrac{\% \text{ water-dispersible material}<60\mu m}{\% \text{ material}<60\mu m} \times 100$

Tab. 2: *Analyses of representative samples from forested soils near Gerona.*

samples 7, 8, 9 and 10 are clearly more stable than samples 2, 5, 11 and 13. In most cases, as expected, the A horizons are more stable than B horizons although this is not always so. Reasons for this probably include the fact that organic substances which promote stabilisation in these soils are sufficiently present in both horizons. Also, at some sites the A horizons could have in the recent past been disturbed either by fire or by the harvesting of undergrowth.

Most of the data plotted in fig.4 also suggests that it is the fraction finer than 10 μm which differs most between the two analyses and that this is usually incorporated in relatively fine water stable aggregates. This is again true for the material from samples 18 and 19, containing illuviated clay, but in this case all size classes contain water stable aggregates. The coarser water stable material is, for most samples obviously, composed of silt grains. The high water stability of the clay and very fine silt fraction is most pronounced in samples 24, 25 and 26 (fig.5) where between only 9 and 24% of the high (35%) amount of clay present is not water stable. The dispersion of the deeper horizons can be explained by the high percentage of exchangeable sodium and the low electrolyte concentration (tab.2). The percentage of the clay fraction that is wa-

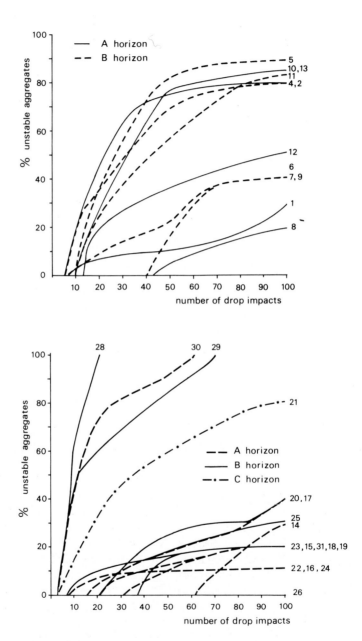

Fig. 6: *The stability of 4–5 mm aggregates as indicated by the water-drop test.*
The lines show the cumulative percentage of the aggregate population destroyed after a given number of drop impacts.

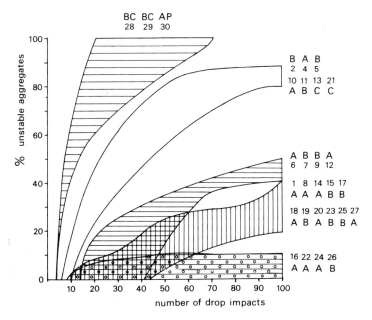

Fig. 7: *"Stability" fields within which the lines plotted in fig.6 are clustered.* The samples occurring in each field are indicated, together with the soil horizons from which they originate.

ter dispersible (tab.2) was found to vary from 8 to 51%, with most samples having values of about 20%. There is no clearly consistent difference between A and B horizons, nor a clear relationship with free iron or organic carbon contents.

Aggregate stability determinations indicating the relative stability of 4–4.8 mm aggregates are plotted in fig.6 where the soil horizons are also indicated. The cumulative relationships tend to cluster in particular fields. Samples 28 to 30 are the most unstable, in the case of samples 28 and 29 this reflects the water soluble salt content and in the case of sample 30 the effects of cultivation. The fields in which the cumulative relationships appear to fall are shown in fig.7. Apart from the unstable field mentioned above four other fields can be observed. The most stable of these are occupied mainly by samples from A horizons, or by B horizons with high clay contents. Only two are Bw horizons on granodiorite which generally occupy the more unstable fields. The two relatively stable Bw horizons (15 and 17) on granodiorite contain relatively low amounts of water dispersible clay and high amounts of organic carbon (tab.2). Two A horizons occupy relatively unstable fields (4 and 10). Although the organic carbon content of sample 10 is low, this is not the case for sample 4. Sample 4 is a very sandy sample (79 percent sand) from beneath *P. pinar* and the organic carbon may not be present in a form conducive to stability. It can further be seen from fig.4 that these samples contain very little water stable silt or clay.

A complete analysis and explanation of microaggregation and soil property relationships, based on the data presented above, is beyond the scope of this paper. The results are described only in order to illustrate that a comparison of standard grain size analysis with the results of water dispersible silt or clay determinations provides information on the microaggregation of the soil which can be useful in terms of understanding its reaction to rainfall.

4 Conclusion

The three examples described above illustrate how the behaviour of contrasted soils from three locations in Spain is either influenced by or related to, the aggregation of the silt and clay fraction. This was quite simple to investigate using slightly modified standard analytical procedures which may be used in combination with thin sections.

In the case of the highly calcareous soils, the water stable silt fractions at the first location consisted mainly of calcium carbonate particles and not of microaggregates, although the latter were present. Weak aggregates at the second site were easily broken down in the absence of organic material. Raindrop action resulted in the segregation of coarser and finer particles (primary or secondary) and in compaction.

For the soils on granodiorite, differences in the aggregation of the fine silt fractions were most apparent. These were related to organic matter and clay content and appear to be associated with differences in erodibility.

References

CHITTLEBOROUGH, D.J. (1982): Effects of the method of dispersion on the yield of clay and fine clay. Australian Journal of Soil Research, **20**, 339–346.

FINLAYSON, B.L., GERITS, J. & VAN WESEMAEL, B. (1987): Crusted microtopography on badland slopes in southeast Spain. CATENA, **14**, 131–144.

GERITS, J.J. (1989): Physico-chemical thresholds for sediment environment and transport. Ph.D. thesis, University of Amsterdam.

GERITS, J., IMESON, A.C., VERSTRATEN, J.M. & BRYAN, R.B. (1987): Rill development and badland regolith properties. CATENA SUPPLEMENT **8**, 141–160.

IMESON, A.C. & VERSTRATEN, J.M. (1985): The erodibility by ultrasonic dispersion and waterdrop impact. Geoderma, **34**, 185–200.

IMESON, A.C. & VERSTRATEN, J.M. (1988): Rills on badland slopes: a physico-chemically controlled phenomenon. CATENA SUPPLEMENT **12**, 139–150.

IMESON, A.C. & VIS, M. (1984): Assessing soil aggregate stability by ultrasonic dispersion and water-drop-impact. Geoderma, **34**, 185–200.

MÜCHER, H.M. & DE PLOEY, J. (1984): Formation of afterflow silt loam deposits and structural modification due to drying under warm conditions: an experimental and micromorphological approach. Earth Surface Processes and Landforms, **9**, 533–531.

SEVINK, J. (1988): Soil organic horizons of Mediterranean forest soils in northeast Catalonia (Spain): their characteristics and significance for hillslope runoff and effect of management. CATENA SUPPLEMENT **12**, 31–43.

SHANMUGANAIHAN, R.T. & OAPES, J.M. (1982): Effect of dispersible clay on the physical properties of the B horizon of a redbrown earth. Australian Journal of Soil Research, **20**, 315–324.

Address of authors:
A.C. Imeson & J.M. Verstraten
Laboratory of Physical Geography and Soil Science
University of Amsterdam
Dapperstraat 115
1093 BS Amsterdam
The Netherlands

SOILS FROM LIMESTONE AND DOLOMITE IN THE CENTRAL ALGARVE (PORTUGAL), THEIR QUALITIES IN RESPECT TO GROUNDWATER RECHARGE, RUNOFF, ERODIBILITY AND PRESENT EROSION

R. **Jahn**, D. **Pfannschmidt** & K. **Stahr**, Stuttgart

Summary

A Mediterranean karstic landscape in the central Algarve (Portugal) was studied with regard to water regime and erosion. The distribution of soil units is related mainly to the petrography of rocks and relief. The dominant Regosols, Chromic Cambisols and Chromic Luvisols are formed by carbonate dissolution, rubefication, clay illuviation and ped formation. During the rainy season in winter, a soil of 50 cm depths needs about 150–200 mm of precipitation in order to reach field capacity (pF 1.8); groundwater recharge for an average year ($\approx$ 600 mm rainfall) amounts to 150 mm. In general, groundwater recharge is mainly affected by the distribution and amount of precipitation, soil type, as well as soil depth. The well-aggregated soils show initial runoff and erosion when the volume of coarse pores (air capacity) of the topsoil ($\approx$ 20–30 cm deep) is filled by a single rainfall event. Under present conditions, erosion only occurs in the steep parts of slopes having shallow soils. The total soil loss for the computed year 1984/85 originates from only 30% of the investigated catchment. Erosion losses for 6% of the area was as high as 750 kg ha^{-1} year^{-1} (54 μm year^{-1}), whereas 15% of the area lost 450 kg ha^{-1} year^{-1} (32 μm year^{-1}). The new formation of soils from limestones is estimated to be 1 μm year^{-1}. The denudation rates explain the degradation of soils in historic times. After 1000 to 10000 years the complete loss of these slope-soils can be expected. The soils of the plains are normally not affected by erosion and are not endangered.

Zusammenfassung

Ein Mediterranes Karstgebiet im mittleren Algarve (Portugal) wurde hinsichtlich seines Wasserhaushaltes und seiner Erosion untersucht. Die Verbreitung der Böden ist stark an die Petrographie des anstehenden Gesteines und an das Relief gebunden. Für die Regosole, Chromic Cambisole und Chromic Luvisole sind Karbonatlösung, Rubefizierung, Tonverlagerung und Aggregierung die dominierenden Prozesse. Bei durchschnittlicher Bodentiefe von 50 cm werden etwa 150–200 mm Nieder-

ISSN 0722-0723
ISBN 3-923381-17-4
©1989 by CATENA VERLAG,
D-3302 Cremlingen-Destedt, W. Germany
3-923381-17-4/89/5011851/US$ 2.00 + 0.25

schlag benötigt um im Winter die Feldkapazität (pF 1,8) zu erreichen. In diesem Falle beträgt die Grundwasserneubildung bei 600 mm durchschnittlichem Niederschlag etwa 150 mm. Sie wird stark von Verlauf und Höhe der Winterniederschläge, des Bodentyps und der Entwicklungstiefe der Böden beeinflußt. Bei den gut aggregierten Böden tritt Oberflächenabfluß und Erosion erst auf wenn das Grobporenvolumen (Luftkapazität) der Oberböden ($\approx$ 20–30 cm mächtig) gefüllt ist. Unter den gegenwärtigen Bedingungen tritt Erosion nur in den steileren und flachgründigen Bereichen der Hänge auf. Der Gesamtabtrag im Meßjahr 1984/85 stammte von nur 30% der Fläche des untersuchten Wassereinzugsgebietes. Auf 6% der Fläche wurde ein Abtrag von 750 kg ha^{-1} year^{-1} (54 μm year^{-1}), auf weiteren 15% der Fläche 450 kg ha^{-1} year^{-1} (32 μm year^{-1}) festgestellt. Die Bodenneubildung aus Kalkstein dagegen wurde mit ca. 1 μm year^{-1} errechnet. Die festgestellten Abtragsraten erklären die Bodendegradation der Hänge in historischer Zeit und lassen in einem Zeitraum von 1000–10000 Jahren den totalen Verlust dieser Böden befürchten. Die Böden der Ebene sind dagegen von Erosion unbeeinflußt und nicht gefährdet.

1 Introduction

The group of Mediterranean soils deriving from limestones show great variability with respect to their physical and chemical properties (CARDOSO 1965, BESSA 1966, KOYUMDJISKY et al. 1966, VERHEYE 1973, WOERNER 1987). The variation of rock and soil properties as well as the Mediterranean climate results in great differences in the water budget of any selected site.

Considering the great variability of soils, as well as the high costs of long-term investigations, both water budget and soil erosion studies are very scarce for Portuguese soils. The application of improved techniques in providing more groundwater for irrigation, even in places with limited occurrence of groundwater, regional water budgets have recently become unbalanced. In order to investigate the present situation in a selected limestone area of Portugal and to give recommendations for the future, an interdisciplinary research project was founded by the Technische Universität Berlin in 1983, integrating hydrologists, agriculturists, soil scientists and sociologists. The project area covers part of the plain of Faro and the adjacent hilly region to the north. The entire area has developed from Jurassic limestones, dolomites and marls.

The purpose of this paper is to explain physical properties of soils derived from limestones, marls and dolomites and their qualities with respect to the water budget as well as the present soil loss of the selected area.

2 Methods

An area of about 2100 ha was mapped in detail according to the FAO classification system (FAO-UNESCO 1974) using fieldchecks and air photographs. A total of 14 profiles, giving 48 horizons, were analysed in detail.

- Water regime analyses were carried out by measuring the moisture with tensiometers (wet conditions) and gypsum electrodes (moist conditions) for 5 selected sites.

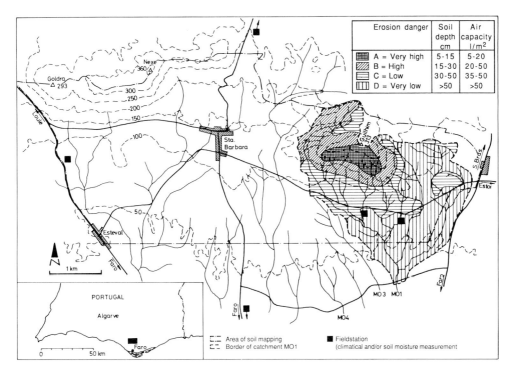

Fig. 1: *Potential erosion endangered areas of the study area. (Guilhim hill, Sta. Barbara, Algarve)*

- Particle size analysis was determined according to SCHLICHTING & BLUME (1966), after H_2O_2 and HCl treatment as well as without carbonate dissolution.

- Water retention was determined with pressure plates, using undisturbed samples (100 cm^3 cores) for retention <3 bar, and disturbed samples for retention 3–15 bar.

- Saturated water infiltration was measured in the field using double-ring infiltrometers according to DNA (1972).

- Potentiometric analyses of pH according to SCHLICHTING & BLUME (1966).

- Conductometric analyses of total carbon by oxidation of the samples at 1000°C and of carbonates after digestion with H_3PO_4 at 80°C using a WÖSTHOFF apparatus.

- Analysis of dolomite by HCl extraction of Ca and Mg at 100°C.

- Analysis of cation exchange capacity with Na acetate at pH 8.2 and of exchangeable cations with NH_4 acetate at pH 7 according to BOUWER et al. (1952).

- Analysis of electrical conductivity in a 1:2.5 soil:water suspension and computing the values at field capacity ($\approx$ saturation extract).

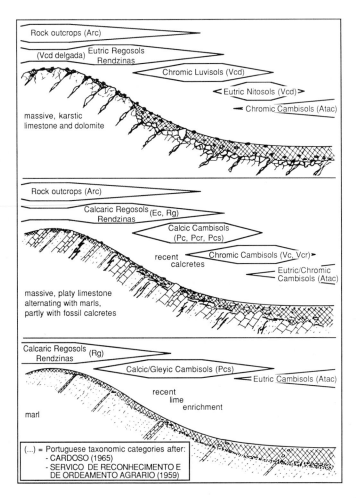

Fig. 2: *Occurrence of soils dependent on rock facies and relief position in the jurassic limestone area of Algarve.*

- Runoff data were collected by KLETSCHKE (1988) over 4 years.

- Sediment load of runoff was estimated from 29 samples collected at different runoff events and values. For each sample, 5 bottles (2 l each) of suspended sediments were collected at the middle of the channels.

3 The research area

Fig.1 shows the location of the project area. The shadowed area marks a small catchment where the present erosion was estimated. The project area covers some 30 km^2 and consists of a part of the plain of Faro (50 m a.s.l.), of foothills, and of hills up to 360 m a.s.l.

The average annual precipitation is about 600 mm, ranging from 200 mm to

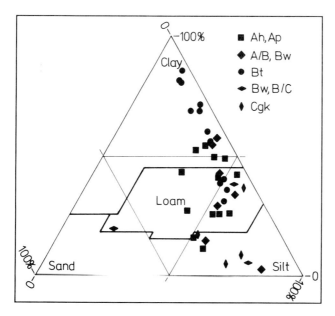

Fig. 3: *Texture of soils derived from limestones, dolomites and marls in the Algarve (data based on fine earth <2 mm, carbonates have not been removed).*

more than 1000 mm, whereas the potential evapotranspiration is about 1600 mm (FARIA et al. 1981). The soils in the research area are mainly found on Jurassic (Oxford, Kimmeridge, Portland) limestones, dolomites and marls. Only a very small part of the area is covered by Pleistocene quartzitic sediments (SERVIÇOS GEOLOGICOS DE PORTUGAL 1985).

The plains are used intensively for vegetable and fruit production. In the foothill region traditional rainfed fruit-culture with irregular soil treatment is characteristic, whereas the hill region is mostly covered by different stages of garigue. From old abandoned terraces, a more intensive landuse from past centuries is evident, even in the hill region.

4 Soils and soil properties

4.1 Soils

The occurrence of soils depends mainly on the facies of the carbonatic parent material, as shown in fig.2. Whereas the bedrock in the research area has, more or less, similar chemical and mineralogical properties, the differences of the physical properties in connection with the slope position can best explain the existing variability of the soils. The most important soil-forming processes are carbonate dissolution, rubefication, clay illuviation and formation of peds (JAHN et al. 1988).

Depending on the degree of erosion, fairly old and well developed Chromic Luvisols also defined as Terra Rossa are found in areas of massive limestones and dolomites with deep karst clefts and with rock outcrops limiting erosion. From

Depth cm	Color	Peds	Skeleton >2 mm %	Sand 2–.63 mm %	Silt .63–.002 mm %	Clay <.002 mm %	Bulk density kg·dm^{-3}	Content of pores >50 μm %	Content of pores 2–50 μm %	Content of pores <2 μm %
Eutric Nitosol (Rod) on dolomite, rainfed olive plantation										
Ap 0–22	5 YR 3.5/3	sub-angular	6	23.5	32.9	43.6	1.31	21.5	11.6	19.8
Bt1 22–40	2.5 YR 3.5/5	angular	<1	7.9	22.9	69.2	1.31	14.6	9.2	29.1
Bt2 40–70	2.5 YR 3.5/5	angular	<1	5.4	12.7	81.9	1.41	6.7	8.0	34.6
Bt3 70–100	2.5 YR 3.5/5	angular	<1	5.2	14.0	80.8	1.33	14.3	7.9	30.0
Bt4 100–150	2.5 YR 3.5/5	angular	<1	2.8	11.7	85.5	1.31	11.6	9.5	31.8
R/Bt 150–160	2.5 YR 3.5/5	angular	13	0.8	17.5	81.8	—	—	—	—
Chromic Luvisol (Min) on dolomite, irrigated citrus plantation										
Ah 0–8	2.5 YR 3/4	sub-angular	2	16.0	58.5	25.5	1.36	17.8	11.4	21.9
Bw 8–25	2.5 YR 3/4	sub-angular	8	8.6	57.3	34.1	1.49	13.0	9.9	23.5
Bt 25–40	2.5 YR 3/4	angular	13	11.5	49.1	39.4	1.44	14.1	10.9	23.2
R/Bt 40–60	2.5 YR 3/4	angular	≈90	9.9	49.0	41.1	—	—	—	—
Gleyic Cambisol (Rato) on marl, rainfed carob plantation										
Ap 0–17	10 YR 6/4	sub-angular	6	13.8	60.5	25.7	1.35	12.4	19.4	16.0
Bw 17–38	10 YR 6/4	sub-angular	6	13.9	83.6	2.6	1.35	13.9	7.2	24.7
B/C 38–52	10 YR 6/4	subangular to structureless	19	17.2	77.5	5.4	1.33	12.1	10.2	23.4
Cgk 52–90	10 YR 6/4 2.5 Y 7/3 5 Y 7/2	structureless	12	26.1	68.9	5.0	1.31	10.2	13.3	22.2

Tab. 1: *Profile description of three typical soils found on dolomite and marl in the Algarve.*

	pH (CaCl$_2$)	Organic matter %	MgCa(CO$_3$) %	CaCO$_3$ %	CEC meq kg^{-3}	Ca %	Mg %	K %	Na %	Σ %	Elec. Cond. at pF 1.8 mS cm^{-1}
Eutric Nitosol (Rod) on dolomite, rainfed olive plantation											
Ap	7.7	1.6	—	0.7	176	88	6	6.0	0.9	100	1.2
Bt1	7.6	0.9	—	0.3	208	86	10	3.4	0.7	100	1.0
Bt2	7.6	0.6	—	0.1	214	79	17	3.2	0.7	100	0.9
Bt3	6.1	0.4	—	0.1	224	63	24	3.3	0.9	91	0.7
Bt4	6.0	0.4	—	0.1	247	48	26	3.4	0.7	78	1.1
R/Bt	7.5	0.2	29.1	1.0	194	61	35	3.5	1.0	100	0.9
Chromic Luvisol (Min) on dolomite, irrigated citrus plantation											
Ah	7.2	3.0	2.2	0.0	275	57	33	8.8	1.2	100	2.2
Bw	7.4	1.7	2.6	0.0	260	63	32	3.0	2.6	100	1.5
Bt	7.5	1.5	3.1	0.0	247	50	44	2.5	3.8	100	1.7
R/Bt	7.6	1.5	10.6	0.0	244	42	52	2.3	4.0	100	2.1
Gleyic Cambisol (Rato) on marl, rainfed carob plantation											
Ap	7.4	2.0	—	69.6	212	93	3.7	2.0	1.3	100	0.7
Bw	7.6	1.6	—	68.2	210	94	3.5	1.2	1.2	100	0.7
B/C	7.7	1.4	—	70.3	196	94	3.9	1.2	1.1	100	0.7
Cgk	7.8	0.2	—	77.8	122	93	4.6	0.5	1.8	100	0.4

Tab. 1: *Continuation*.

the weakest and most unstable material, silty to clayey marl, Regosols and shallow Cambisols have developed recently (STAHR et al. 1984). Profile data of three typical soils, found on dolomite and marl, are presented in tab.1.

4.2 Physical and chemical properties

As a typical feature of limestone-soils, almost all have loamy to clayey texture (fig.3, tab.1). The soils found on massive limestone and dolomite are mostly heavy clayey soils with strongly aggregated polyedric peds. A greater variability of texture is found in soils derived from marls due to their weathering stage (degree of decarbonatization). High contents of silt were found in soils with a low weathering stage on marl. In contrast, Bt-horizons of soils derived from limestone or dolomite may have clay contents up to 85%. The A-horizons in general contain additional sand and silt, either by loss of clay due to illuviation, selective erosion of clay and silt or incorporation of silty eolian dust.

The high clay content, formation of peds by swelling and shrinking, high bulk density and micro-aggregation by oxides are the main factors affecting the pore size distribution of these Mediterranean soils; the pore sizes vary within a wide range, depending on the development of typical horizons (fig.4).

The general characteristics of pore size distribution are:

- a relatively high amount of coarse pores (more than 5%), which enables water infiltration, even of clayey Bt-horizons,

- the pores retaining plant-available water are present to a considerable level (more than 10%) only in silty soils derived from marl, but are limited (less than 10%) within the Bt-horizons of Luvisols,

- because of the high clay content, the amount of water held at pF 4.2 is generally high (more than 20%), especially within Bt-horizons.

The water infiltration in A-horizons shows rates ranging from 60–1200 cm day^{-1}. The variability depends to a great extent on landuse. Lower infiltration rates are found in areas with macchia or garigue, whereas higher rates are found in fruit plantations where loosening of the soil is frequent. The infiltration rates are in most cases much higher than the measured rainfall intensities. In the Bt-horizons, rates <3 cm day^{-1} were measured. This means that even in the heavy clayey Bt-horizons seepage is possible, but that during heavy rain they will impede drainage.

These results were obtained through field observation, which indicated the necessity for heavy rain exceeding 20 mm to produce runoff, even for moist soils. This is the lower limit of air capacity of topsoils.

Due to the presence of carbonates within the area of Jurassic sediments, strongly leached soils do not occur. In addition, in fairly old decalcified soils, a base saturation less than 80% was not observed. Normally, the exchange complex is still dominated by Ca, but the Mg saturation increases remarkably with the presence of dolomite (tab.1). The exchange capacity with 200 to 300 meq kg^{-1} (fine earth) is fairly high in Nitosols and Luvisols as well as in Cambisols.

In general, a pronounced enrichment of H_2O soluble salts could not be found the whole year round. In irrigated plantations on well drained soils, using ir-

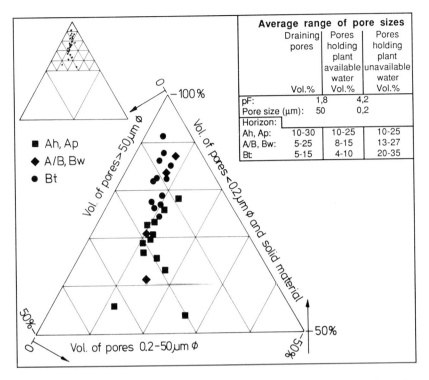

Fig. 4: *Pores size distribution of soils from limestones, dolomites and marls in the Algarve, Portugal.*

rigation water having an electrical conductivity below 1 mS cm^{-1}, the sum of soluble cations reached 10 meq kg^{-1} (fine earth) as a maximum. The content of organic matter within irrigated citrus plantations and rainfed fruit plantations in the plains is medium to low, whereas on slopes with decreasing landuse intensity and increasing biomass of garigue and macchia (also with cooler temperatures), the content of organic matter increases.

4.3 Soil moisture and water balance

Fig.5 shows the change of moisture, expressed as pF-values, from autumn 1984 to summer 1986 in the top- and subsoil of an Eutric Nitosol under rainfed management. In addition, daily values of potential evapotranspiration and precipitation are shown.

An important finding is that a minimum rainfall of about 300 mm is necessary to moisten (pF <1.8) the soil to a depth of 150 cm. The more common soil depths in the research area, of about 50 cm, were moistened after 150–200 mm rainfall. These values are also dependent on the amount of rainfall as well as on the interval between rainstorms.

Furthermore, it was observed that single rainstorms of more than 20–30 mm day^{-1} are necessary to produce runoff. Much smaller amounts have been recorded for some neighbouring watersheds which have soils derived from Paleozoic rocks (KLETSCHKE 1988). For

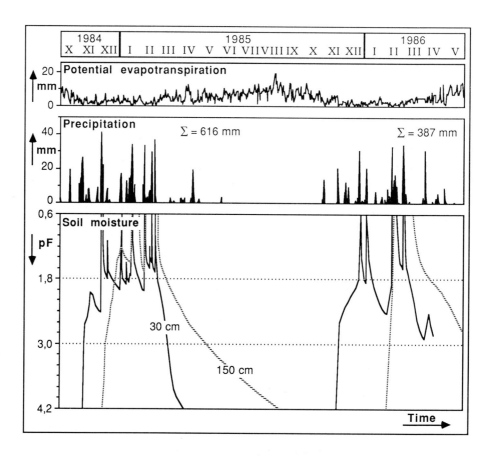

Fig. 5: *Water tension, precipitation and potential evapotranspiration for an Eutric Nitosol near Sta. Barbara, Algarve, under rainfed management from 1984–1986.*

Soil depth	Days with pF <1.8		Rainfall mm		Potential evapotranspiration mm		Seepage mm	
	84/85	85/86	84/85	85/86	84/85	85/86	84/85	85/86
30 cm	55	44	382	242	221	168	162	74
50 cm	71	55	372	232	227	186	146	45
70 cm	86	52	362	211	238	177	125	34
100 cm	91	42	312	139	213	130	99	9
150 cm	78	56	302	164	203	200	99	0

Total rainfall 1984/85: 616 mm, 1985/86: 387 mm

Tab. 2: *Days with possible seepage, precipitation, potential evapotranspiration and estimated seepage for different depths of limestone soils in the Algarve, Portugal.*

the initiation of runoff, previous rainfall is necessary in order to moisten the soils. No hydrophobic surfaces were observed.

Between rainstorms (dry period 1–10 days) the pF increases to values around pF 1.8 (fig.5). Accordingly the amount of coarse pores (>50 μm m) of the top-soils having high water conductivity can serve as a rough threshold-value for runoff-forming rainstorms. Impermeable topsoil crusts were not often observed due to the structure of the heavy clayey soils.

As can be seen in fig.5, heavy rain normally stops at the end of February, followed by a quick drying-out of the topsoils. Only in the deeper parts were moist conditions found to last longer. However by September/October the entire soil down to 150 cm was dry (pF >4.2). This moisture regime is typically xeric.

Using soil moisture, the possible groundwater recharge was estimated. This was achieved by computing the time interval during which the soil moisture fell below pF 1.8. During this time, the coarse pores were filled with water and seepage was possible. All rainfall during this time, minus the evapotranspiration, was calculated as groundwater recharge. The results shown in tab.2 are based on the measurements done on an Eutric Nitosol during two rainy seasons. In general, these results ar only applicable to strongly aggregated soils, commonly Chromic Luvisols and Nitosols. Measurements on different sites having different soil depths did not reveal any great differences at comparable depths. Therefore the different depth studied in the Nitosol can be used for comparison with the same depths of other sites, the error being reasonably small.

5 Soil erosion

Erosion processes removing soil material in the landscape, are still active. The dominant part of the process of removal is due to surface runoff. In order to estimate the present erosion in the area, we investigated a catchment (640 ha) situated on the south slope of the Guilhim hill and on the planes in front of it (fig.1).

5.1 Erosion susceptibility

5.1.1 Rainfall

The precipitation during the rainy season studied in the winter of 1984/85 was 616 mm, and is representative of an average year. During 217 days in the rainy season October 1984 – April 1985, rain was registered on 80 days, whereas 73% of the total rain occurred on 28% (22 days) of the rainy days, with daily precipitation values >10 mm. The highest precipitation for a single rainstorm was registered on February 25, 1985 when 36.6 mm fell within 2 hours. Rainfall data analysis, based on 2-hour values, show the following charactistics (tab.3):

The infiltration rate of the soils at field capacity is mainly greater than 100 mm $\cdot$ h^{-1}. Therefore runoff is rarely caused by high rainfall intensity. Surface runoff is primarily induced when the soil is saturated, as only in this case is the infiltration rate less than rainfall intensity.

5.1.2 Erodibility of soils

Field observations of erosion damage suggest great variability in soil erodibility, which is in fact reflected in the K-factors calculated according to WISCHMEIER & SMITH (1978). Tab.4 shows the K-values of different topsoils, derived from different parent materials.

2-hr rain range (mm):	0–0.5	0.5–1	1–2	2–5	5–10	10–20	>20	Σ
Σ (mm):	28.9	35.2	77.0	194.9	151.8	63.0	64.9	615.7
% of Σ :	4.7	5.7	12.5	31.7	24.7	10.2	10.5	100
cumulative %:	100	95	90	77	45	21	11	
No. of events:	133	46	52	61	22	5	2	321

Tab. 3: *Precipitation characteristics (October – April, 1985) based on 2-hour values.*

Nr.	Soil unit	Parent material	Clay %	Silt %	Sand %	OM %	K* factor	Erodibility class**
VII	Eutric Regosol (chromic)	Dolomite	53	33	14	2.4	0.10	1 very low
II	Chromic Luvisol	Limestone	50	46	5	5.2	0.10	1 very low
VI	Chromic Luvisol	Dolomite	55	36	9	1.5	0.11	2 low
I	Eutric Regosol (chromic)	Limestone	42	54	4	6.5	0.11	2 low
Moi.2	Gleyic Acrisol	Quartzy deposits	3	12	86	2.5	0.14	2 low
Rod	Eutric Nitosol	Dolomite	44	33	24	1.6	0.14	2 low
VIII	Chromic Luvisol	Dolomite	50	42	8	1.4	0.14	2 low
III.1	Gleyic Cambisol	Soil Colluvium/Marl	41	48	12	3.3	0.16	3 low
Moi.1	Orthic Luvisol	Limestone	15	21	64	1.3	0.16	3 low
Alp.3	Eutric Cambisol	Basalt	16	21	64	1.1	0.18	4 low
III.3	Calcaric Regosol (chromic)	Marl	27	43	30	1.6	0.23	5 moderate
Moi.3	Orthic Luvisol	Silt stone	23	29	48	1.0	0.24	5 moderate
Alp.1	Dystric Cambisol	Clay schist	18	41	41	1.9	0.25	6 moderate
Alp.2	Chromic Luvisol	Sand stone	17	27	56	1.1	0.25	6 moderate
Hü	Calcaric Regosol	Pyroclastics	12	50	38	3.7	0.28	6 moderate
St.B.	Calcic Cambisol	Marly clastics	12	57	31	4.5	0.29	7 moderate
Min	Chromic Luvisol	Dolomite	26	59	16	3.0	0.29	7 moderate
Mol	Chromic Luvisol	Dolomite	26	56	18	2.0	0.30	7 moderate
III.2	Calcaric Regosol (light)	Marl	16	51	33	1.8	0.32	7 moderate
Rato	Gleyic Cambisol	Marl	25	64	11	2.0	0.34	8 mod. high

* First approximation of K-value (WISCHMEIER & SMITH 1978)
** Classified after DANGLER & EL-SWAIFY (1976)

Tab. 4: *Texture, organic matter and erodibility of topsoils from different parent materials.*

In addition to soil erodibility being affected by texture and organic matter, the stone content and stability of peds influences soil erodibility to a great extent. Both high stone coverage and high ped stability can generally be found in the red (redder than 7.5 YR) clayey soils found on massive limestones and dolomites. On the other hand, frequent occurrence of rock outcrops at slope positions may limit water retention at these sites.

Measurements of ped stability (HARTGE 1971) show great variability and sometimes only medium stability (Δ GMD = 0.2–3 mm), especially for Bt-horizons. This can be explained by the method, which does not reflect stable micropeds <1 mm. Observations in the field show translocated peds sedimented in the upper part of the catchments. From these observations we assume that the peds are transported as such at the peaks of runoff.

Class	Inclination %	Percent of total area (640 ha) %
0	<2	13
1	2–5	28
2	5–9	13
3	9–12	8
4	12–18	12
5	18–36	22
6	>36	4

Tab. 5: *Distribution of slope classes of the study area, Algarve, Portugal.*

Landuse	runoff-probability
Frequently-tilled Cambisols and Luvisols common under citrus plantations and rainfed fruit tree plantations	low ↑
Macchia and garigue on Cambisols, Luvisols, Rendzinas and Regosols with soil depth >30 cm and lacking rock outcrops	
Rainfed fruit tree plantations on Chromic/Eutric Cambisols without tillage	
Rainfed fruit tree plantations on Luvisols and Gleyic or Calcic Cambisols without tillage	
Macchia and garigue on shallow soils (<30 cm) and many rock outcrops	↓
Sealed areas (roads, housing areas)	high

Tab. 6: *Landuse and runoff-probability in the study area, Algarve, Portugal.*

5.1.3 Slope steepness

The slopes were divided into 7 classes of steepness, in a decreasing sequence from 6 to 0, with 6 at the supreme part of the Guilhim and down to 0 in the plains (tab.5). In class 6, there are also slope angles of up to 90°.

5.1.4 Landuse

At the steepest slope, with angles <36%, macchia are predominant. Slopes having angles from 13–36% had mostly extensive rainfed farming (partly terraced), where the macchia have spread again after farming had ceased. With decreasing slope angle, rainfed farming with frequent tillage becomes widespread. Tillage results in the loosening of the topsoil, once per year after the first rain. In the flat of the plain, extensive rainfed farming is absent. Only the intensive forms occur along with irrigated fruit and vegetable farming. Tab.6 shows the general succession of landuse in respect to the probability of runoff and consequently of soil erosion can be found, taking soil types, soil depth, amount of coarse pores, infiltration rates and landuse into account.

5.2 Potential erosion endangered areas

Because of great variability within the investigated landscape, the different erosion risks should be known. As basic information we considered the thickness of the soils and their different capacity of water storage. The capacity of water storage of the soil was divided into the total storage capacity of the soil, and the volume of coarse pores, which is able to absorb precipitation at once. The lat-

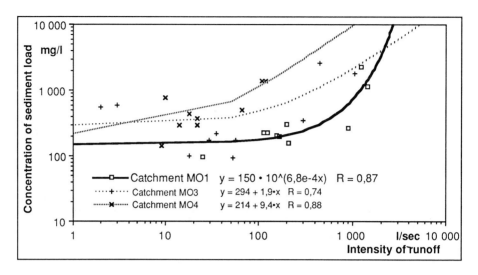

Fig. 6: *Sediment load versus runoff intensity in a catchment near Sta. Barbara, Algarve.*

ter corresponds to the air capacity and was used for the soil layer above the Bt-horizons. The Bt-horizons with a percolation rate of about 1 mm h^{-1} were neglected for the calculation of rapid water absorption.

The air capacity of the soil above the Bt-horizons (except for very shallow soils) is approximately 30 l m^{-2}. This may be a threshold value for precipitation events causing surface runoff, if the soil has a water content at about field capacity.

In the investigated sites, the soil depth generally decreases with increasing slope angle. Shallow soils, greater amounts of bare rock and skeletal soils on the steeper slopes promote runoff. Therefore runoff is first obtained from the steepest slopes of the Guilhim, having soil depths between 5 and 15 cm. These soils only have an average air capacity of about 15 l m^{-2}. On this slope position, erosion takes place more often (area "A" in fig.1).

The next area to produce surface runoff is the upper and middle part of the slope, where soil depths range from 15 to 30 cm. Here we have an air capacity which can absorb rainwater of about 30 l m^{-2}, (area "B" in fig.1). The area having middle to deeply-developed soils on the lower parts of the slope and within the plain, correspond to "C" and "D" on the map.

The lower parts of the slope are protected from upslope by a narrow dendritic network of drainage channels.

5.3 Areas with current formation of runoff

Soil erosion was estimated from the sediments transported by channel flow, calculated by flow data and sediment-load concentrations of samples (fig.6). In order to find the area from which the measured runoff originated, the free pore volume, which can rapidly absorb precipitation, was used as a base value.

To determine the dimension of the area forming runoff, the amount which

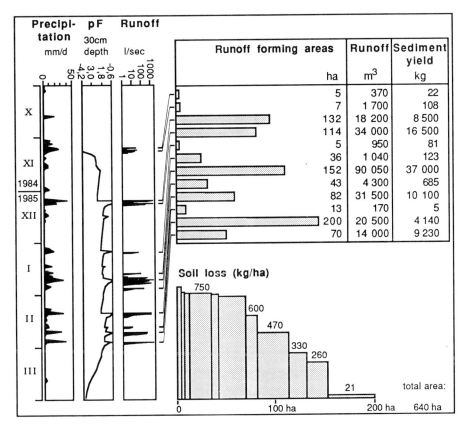

Fig. 7: *Size of runoff forming areas and soil loss (catchment MO1, 1984/85), Sta. Barbara, Algarve.*

is necessary to fill the air-filled pores of the upper part of the soil was subtracted from the precipitation. The size of the area was the result of dividing the discharge of runoff by the residual water from rainfall.

The areas where runoff occurred in 1984/85 are shown in fig.7.

5.4 Runoff and sediment load

Measurements of the sediment concentration at different runoff values generally showed a progressive increase of sediment load with increasing runoff intensity. Fig.6 shows a plot of suspended sediment load versus runoff intensity for the discussed watershed (MO1) and, for comparison, two smaller adjacent watersheds (MO3 and MO4). Suspended sediment concentration for MO1 was 94 mg l^{-1} to 2300 mg l^{-1}, and for Mo3 and MO4 was 13 to 1450 l sec^{-1}. Coarse material moving at the bottom of the channels, was not considered because the observed channels were almost free of bedload.

The three different channels showed great differences in their suspended sediment to runoff intensity ratios, which depended on the distribution of erosion endangered areas within a specific catchment. Furthermore, in addition to runoff intensity, the date of the observed event within the season, time of sampling within one event, length of period since the last event, magnitude of the last event, the total runoff of the observed event and intensity of rainfall all influence the sediment to runoff ratio.

5.5 Actual degree of soil erosion

Provided that the soils of different areas have similar erodibility, the sediment yield of runoff can be directly related to the area producing this runoff. In this case, the sediment yield corresponds to the soil loss of this area, because no transported material is deposited within the watershed. For each rainstorm, fig.7 shows the area which was affected by runoff and subsequently by erosion.

In the rainy season 1984/85 only 200 ha (out of the total 640 ha) of the investigated catchment contributed to the production of overland flow and consequently to runoff yield. This means that the calculated soil loss of about 86 t only arose from about 30% of the catchment area and half of the total soil loss comes from just 10% of the catchment.

At the steepest slope positions (about 6% of the catchment area), which are mostly affected by soil loss, a soil loss of about 750 kg ha^{-1} was calculated (area A in fig.1). For the slopes steeper than 15%, a soil loss of about 450 kg ha^{-1} was calculated (area "B"). Gentle slopes give a negligible rate of soil loss.

For an average bulk density of 1.4 kg dm^{-3} the soil loss was converted into μm year^{-1}. Because the investigated period experienced the average precipitation, and other factors are not in contradiction, the calculated soil loss will be considered as the present erosion rate. Given this assumption, soil loss amounted to 54 μm year^{-1} in the steepest slope positions, with the shallower soils. The soils of this area, with a depth of 5–15 cm, will be entirely removed in about 3000 years, providing the erosion rate remains constant. For slopes exceeding 15% in steepness and having a soil depth of about 30 cm, an erosion rate of 32 μm year^{-1} was calculated. The soils existing in this area today will be removed, according to the calculation above, in about 4500–9000 years. However, the process will be accelerated in the lower slope position if the upper slope is cleared of the soil mantle, thus losing its water storage capacity.

The regeneration rate of soils calculated by possible dissolution of limestone and possible amount of leaching ($\approx$ 150 mm year^{-1} for a 5 dm soil) will be less than 1 μm year^{-1} — much less than the soil loss. As an average 4 to 6 meq Ca + Mg l^{-1} were found in soil solution and groundwater, even additional loess deposits, which were not included in the prediction, would not be able to balance the erosion rate at the slope.

The deep soils in the flat land usually do not produce any runoff. If the rate of rainfall exceeds the rate of infiltration, water will be stored in small surface depressions. However, it is unlikely to flow over due to a fairly good microrelief left by tillage operations, as well as the effect of walls, hedges and ditches in the cultivated land.

If the present circumstances continue, erosion will take place in the future only on shallow soils. However, the shallow soils will become less erodible as more and more stones will be left on the soil surface, thus protecting the soil. The landuse in its present form on the deeper soils in the plain will not be endangered by small soil losses.

6 Conclusions

The site studied can be subdivided into three natural units with respect to soil properties and erosion features.

The first unit comprises the shallow soils on the steeper slopes with macchia and extensive rainfed farming. In this unit (area "A" and "B" in fig.1) soils will degrade further. With the current erosion rate of the steepeest slopes, the soil thickness will decrease in about 500 years by 2.5 cm. Therefore existing sites with shallow soils of 5 cm, already abundant on the upper slopes of the Guilhim, will be reduced by one-half. By the same process, water storage will be restricted. Therefore, the aridity will increase during summer. Assuming that these soils have already been used by man for over 3000 years, and that the erosion rates are as calculated, it is likely that they had an average thickness of 20–30 cm and supported a natural forest.

The second unit consists of sites, used today for extensive rainfed farming with irregular soil tillage, being transformed more and more into an area with intensive soil management. The lower slopes are especially being used for intensive landuse. This will cause an increase in soil tillage, and thus erosion will be high for the first years after cultivation (area "C" in fig.1).

The third unit (area "D" in fig.1) reflects the deeper soils of the plain. It is the most resistant unit against erosion for present landuse.

In order to limit the degradation of the soils of the first unit, a reduction of sheep grazing is recommended such that the spontaneous macchia vegetation can increase in density. An intensification of rainfed farming should not be proposed.

Intensified landuse in the second unit (area "B") can only be recommended if measures are used to prevent erosion, such as terrace rebuilding, contour ploughing, use of cover crops or mulch.

Acknowledgements

This research was part of the interdisciplinary project "Artificial Groundwater Recharge in Semiarid Climates — Effects on Agricultural Structures, Soil Properties and Socio-economic Significance", financed by the Technische Universität Berlin. The authors thank all members of the research project for helping with analyses and field measurements. Special thanks go to Mr. Th. Kletschke for providing runoff data, and also to the farmers who accepted measurements on their farms with great endurance.

References

BESSA, M.R.T. (1966): Sols Mediterraneens de Portugal meridional. Transactions of Conference on Mediterranean Soils, Madrid, 1966, 173–186.

BOUWER, C.A., REITEMEIER, R.F. & FIREMAN, R. (1952): Exchangeable cation analysis of saline and alkali soils. Soil Sci., **73**, 251–261.

CARDOSO, J.C. (1965): Os Solos de Portugal, 1-A Sul do Rio Tejo. Lisboa.

DANGLER, E.W. & EL-SWAIFY, S.A. (1976): Erosion of Hawaii Soils by Simulated Rainfall. Soil Sci. Soc. Am. J. vol. **40**, 769–773.

DNA (1972) FACHNORMENAUSSCHUSS WASSERWESEN IM DEUTSCHEN NORMENAUSSCHUSS (DNA): DIN 19682 Bl. 7: Felduntersuchungen, Bestimmung der Versickerungsintensität mit dem Doppelzylinder-Infiltrometer. Deutscher Normenausschuß, Berlin.

FAO-UNESCO (1974): Soil map of the world. Volume I - Legend. Paris.

FARIA, J.M.R., GODINHO, S., ALMEIDA, M.J.R. & MACHADO, M.S. (1981): O Clima de Portugal. Lisboa.

HARTGE, K.H. (1971): Die physikalische Untersuchung von Böden. Enke, Stuttgart.

JAHN, R., STAHR, K. & LASSONCZYK, B. (Hrsg.) (1988): Verbreitung, Genese und Standorteigenschaften von Böden im Algarve. Landschaftsentwicklung und Umweltforschung, 55, Berlin.

KLETSCHKE, Th. (1988): Beitrag zur Klärung des Gebietswasserhaushaltes des mesokänozoischen Zentralalgarve (Portugal). Dissertation, TU Berlin.

KOYUMDJISKY, H., YAALON, D.H. & DAN, J. (1966): Red and reddish brown Terra Rossa in Israel. Transactions of Conference on Mediterranean Soils. Madrid, 195–201.

SCHLICHTING, E. & BLUME, H.-P. (1966): Bodenkundliches Praktikum. Hamburg, Berlin.

SERVICOS GEOLOGICOS DE PORTUGAL (1985): Carta Geologica de Portugal, Folha 53-A Faro.

SERVICO DE RECONHECIMENTO E DE ORDEAMENTO AGRARIO (1959): Carta dos Solos de Portugal, Folha 53-A Faro.

STAHR, K., GAUER, J. & JAHN, R. (1984): Bodenvergesellschaftung vom Mediterranraum in die Vollwüste. Mitteilgn. Dtsch. Bodenkundl. Gesellsch., 40, Göttingen, 223–230.

VERHEYE, W. (1973): Formation, classification and land evaluation of soils in Mediterranean areas with special reference to Southern Lebanon. State University Ghent (Belgium).

WISCHMEIER, W.H. & SMITH, D.D. (1978): Predicting rainfall erosion losses — a guide to conservation planning. U.S. Department of Agriculture, Agriculture Handbook No. 537, Washington, D.C.

WOERNER, M. (1987): Die bodenphysikalischen Eigenschaften der wichtigsten Böden des Algarve und ihre Eignung zur Bewässerung. Dissertation, Bonn.

Address of authors:
Reinhold Jahn, Detlef Pfannschmidt & Karl Stahr
Department of Soil Science and Ecology
University of Hohenheim
Emil-Wolff-Str. 27
D-7000 Stuttgart 70
West Germany

RELATIONSHIP BETWEEN SOIL PROPERTIES, SOIL MOISTURE, AND VEGETATION ALONG LOESS-COVERED HILLSLOPES, NORTHERN NEGEV, ISRAEL

R. **Kadmon**, A. **Yair** & A. **Danin**, Jerusalem

Summary

A preliminary study of the variation in the vegetation along arid hillslopes, conducted in the northern Negev desert, has shown that there is a gradual increase in environmental aridity on passing from the rocky upper part to the lower soil-covered portion. In order to check the general validity of the initial findings, a second study site was selected. Two opposite slopes, north and south-facing, representative of the area were chosen and a transect running along each of the slopes was studied in detail. The variables considered were the mechanical properties and electrical conductivity of the soil, the variation of soil moisture over time, the floristic composition, and the abundance of the existing plant species. Results obtained support the conclusion, derived from the initial study, that environmental aridity increases in the downslope direction and, in more general terms, that water availability in a non-sandy desert depends mainly on the ratio of bare bedrock to soil cover.

ISSN 0722-0723
ISBN 3-923381-17-4
©1989 by CATENA VERLAG,
D–3302 Cremlingen-Destedt, W. Germany
3-923381-17-4/89/5011851/US$ 2.00 + 0.25

1 Introduction

Studies dealing with the spatial distribution of vegetation in an arid environment usually emphasize the difference between valleys and hillsopes. The valleys are regarded as relatively wet habitats possessing a mesophylic vegetation with the hillslopes as relatively dry habitats with a diffuse vegetation (e.g. COTTLE 1932, HILLEL & TADMOR 1962, TADMOR et al. 1962, ORSHAN 1985, SHMIDA et al. 1986).

The assumption that hillslopes in an arid area represent a dry environment encounters serious difficulties when applied to the northern Negev desert. A study (YAIR & DANIN 1980) of the distribution of plant communities along a slope, in an area where average annual rainfall is 93 mm, showed that an improved water regime, with a high diversity of plant species and a relatively high percentage of Mediterranean species, was characteristic of the upper rocky slope section. In contrast, a low species diversity and a high percentage of Saharo-Arabian species characterized most of the colluvial, soil-mantled, slope section. These findings were fully supported by a quantitative study of the vegeta-

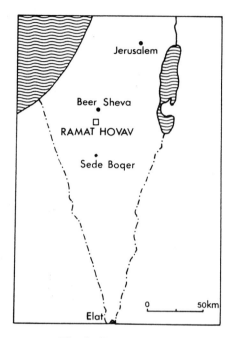

Fig. 1: *Location map.*

tion conducted in the same area, based on Detrended Correspondence Analysis (OLSWIG-WHITTAKER et al. 1983). Such results were explained by the systematic variation in water availability along arid hillslopes.

Detailed hydrological studies conducted at the Sede Boker experiment site (fig.1) clearly indicate that water availability in the study area is mainly controlled by the ratio of bare-bedrock outcrops to soil cover (YAIR 1983, 1987). Where this ratio is high, such as for the uppr rocky slope section, the bare-bedrock outcrops respond quickly to rainfall and frequently generate high runoff yields per unit area. This runoff is immediately absorbed by the soil material, infilling fissures and bedding planes of the bedrock, and by the adjacent soil-covered areas (upper colluvial section). In these environments an improved wa-

ter regime develops. The wetting depths is beyond that which can be attributed to direct rainfall. Even with evaporation a water reservoir is created below 50 cm depth, being available for plants during the long and hot summer season. Conversely over soil-covered areas, due to their high infiltration rate, these areas are capable of absorbing all rainwater of most storms and thus little or no runoff is generated. As rain amounts for most storms are limited, the depth of water infiltration seldom exceeds 30 cm. Most of this water is subsequently lost through evaporation and is therefore of little benefit for biological activity.

The main purpose of this study was to test the hypothesis that the transition from a relatively mesic into a more xerophytic condition, as expressed by soil properties and vegetation identified along the Sede Boker slope, is general and valid at the regional scale.

2 Description of the study area

The area selected, the Hovav Plateau, is located some 35 km north of Sede Boker (fig.1). Average annual rainfall is 170 mm, with mean daily temperatures ranging from 11° in January to 27° in August. Annual evaporation, measured with a class "A" evaporation pan, is in the order of 2100 mm (ZANGVIL & DRUIAN 1983).

The landscape consists of gently sloping hills. As at Sede Boker, the hillslopes are subdivided into an upper rocky, and a lower colluvial, section (fig.2). The rocky slope is composed of very densely jointed flinty Eocene chalk which weathers into cobbles and gravels, embedded in a quite contiguous thin loess veneer. The colluvial slope is composed of a stoneless loess cover that thickens quickly downs-

Relationships between Soil and Vegetation along Hillslopes, Northern Negev 45

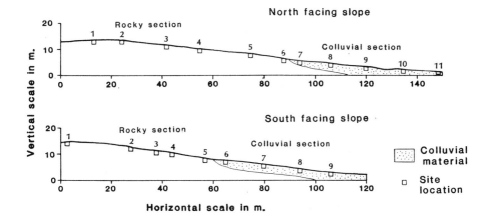

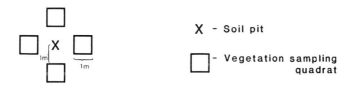

Fig. 2: *a: Slope profiles and site locations; b: Sampling pattern.*

lope. Due to the proximity of the study area to the Haluza dunes, the sand content of the Hovav soil is 40%–60%. The vegetation is composed of sparse semishrubs associated with annual and perennial herbs.

3 Methodology

Two slopes representative of the area were chosen for the study. One is north and the other south facing. A transect running from the top to the base of each slope was then selected with sites for the study of the soil and the vegetation chosen along each transect (fig.2). The distances between sites were sufficiently close to detect any gradual change in the variables studied.

The study of the abiotic environment included the following variables:

a) temporal variations in soil moisture, down to a depth of 40 cm, during the period February–July 1982;

b) particle size composition;

c) electrical conductivity (EC) of soil samples taken from depths of 5–10 cm and 15–20 cm;

d) percent of stones in the soil profile.

The study of the vegetation was based on four distinct quadrats, of one square meter each, surrounding each soil pit (fig.2). The quadrat size was considered suitable being several times greater than

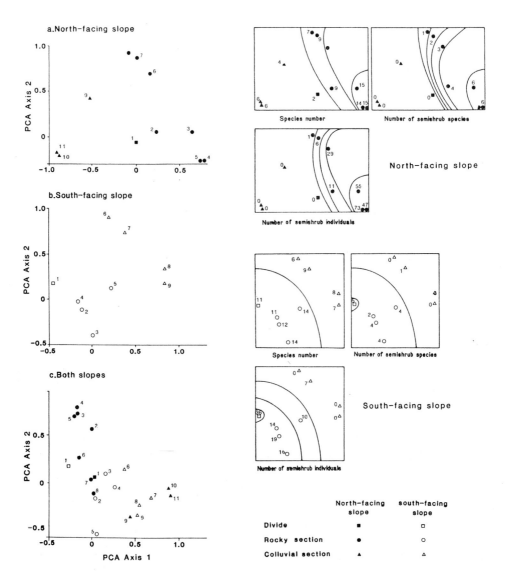

Fig. 3: *PCA results.*

Fig. 4: *Relationships between community characteristics and PCA results.*

the horizontal size of the largest plants. For each quadrat the floristic composition was determined and the number of individual plants belonging to each species was counted. The final list for each site was represented by the sum of its four quadrats.

3.1 Analytical procedures

Soil samples were analyzed through standard laboratory tests. Particle size com-

	N		S		C	
	l	u	l	u	l	u
Ns	5.3	11.7	7.5	12.7	6.1	12.1
Sch	0.0	4.0	0.2	3.5	0.1	3.8
PSch	0.0	32.3	2.7	27.7	1.0	30.4
Ich	0.0	31.7	1.7	14.7	0.6	25.5
PIch	0.0	35.9	1.7	20.7	0.6	30.4
PSa	75.0	45.6	73.7	45.0	74.5	45.4
PIa	98.3	52.6	89.0	66.5	94.9	57.7

N = North-facing slope; S = south-facing slope;
C = combined slopes
l = lower section; u = upper section
Ns = number of species
Sch = number of semishrub species
PSch = percentage of semishrub species
Ich = number of semishrub individuals
PIch = percentage of semishrub individuals
PSa = percentage of annual species
PIa = percentage of annual individuals
(Each value represents a mean)

Tab. 1: *Community characteristics: a comparison between upper and lower slope sections.*

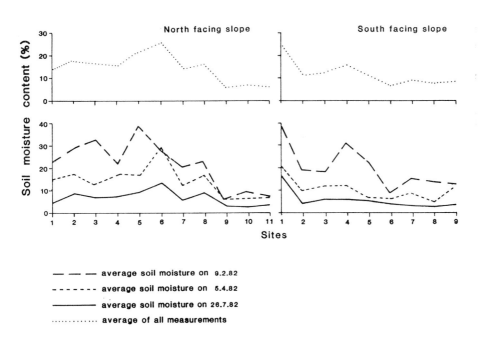

– — — average soil moisture on 9.2.82
- - - - - - average soil moisture on 5.4.82
——— average soil moisture on 26.7.82
.... average of all measurements

Fig. 5: *Spatial and temporal variation in the soil moisture regime.*

position was determined with sieves and the pipette sedimentation method. EC was measured at 25°C with a conductivity meter. Soil moisture was determined gravimetrically with oven dried samples; and the stoniness by visible estimation of the soil profile surface.

The vegetation data were analyzed using Principal Component Analysis (PCA) (NIE et al. 1975). In order to give equal weight to the various stands and compensate for unequal abundance of different life forms, the data was standardized according to COTTAM et al. (1978).

4 Results

4.1 Floristic data

The PCA results are displayed in figs. 3 & 4. Data analysis indicates the following main points:

- A clear-cut separation exists, for each of the slopes considered, between the upper rocky and the lower colluvial slope sections (fig.3).

- The gradients of the communities' characteristics are well reflected in the PCA results. An important feature is the virtual absence of woody plants at the lower part of both slopes and their relative abundance over the rocky slope section (fig.4).

- Despite the diametrically opposite slope aspects, the principle variation is not between slopes but rather along each of the slopes (fig.3, tab.1).

4.2 Abiotic data

4.2.1 Temporal and spatial variation in soil moisture

Soil samples for moisture determination were taken following major rainstorms, and once at mid-summer (fig.5). Data obtained show the following trends:

- The pattern of soil moisture variation along the studied slopes is quite similar for all the dates sampled.

- On the average, as well as for individual dates sampled, the lower part of the colluvial slope (sites 9–11 on the north-facing slope and sites 6–9 on the south-facing slope) displayed the lowest moisture content.

- The highest soil moisture content was always found within the rocky-to-stony slope section as well as at the upper colluvium.

- Extremely low moisture contents (3–4%) were found during mid-summer at the lower colluvial slope sections of both slopes.

4.2.2 Particle size composition and electrical conductivity

The particle size composition, stoniness and EC data obtained for the north-facing slope are presented in tab.2. Sand and silt are the predominant factors. Relatively high clay contents were found at the upper slope (sites 1 & 2) and at the upper colluvium (sites 7 & 8). The soils are quite saline. High figures for the EC were found at a shallow depth of 10–15 cm. As could be expected, salinity is especially high where clay content is high, and low where the sand content is high. Stoniness is high within the rocky

site	depth (cm)	clay	silt	sand	EC	Average stoniness
1	5–10	44.5	42.0	44.5	17.2	2
	15–20	23.5	39.8	36.7	22.9	
	25–30	30.7	47.6	21.7	21.1	
2	5–10	16.7	35.8	47.5	19.4	35
	15–20	30.0	37.3	32.7	25.3	
3	5–10	13.4	34.9	55.7	1.9	39
	15–20	14.8	30.1	55.1	5.0	
4	5–10	12.0	40.2	47.8	5.1	34
	10–15	14.4	36.2	49.4	5.7	
5	5–10	12.1	42.5	45.4	3.4	25
	10–15	16.3	40.7	43.0	10.9	
6	5–10	15.8	43.3	40.9	8.8	30
	10–15	16.9	34.3	48.8	14.5	
7	5–10	14.9	35.7	49.4	3.3	20
	15–20	24.1	41.9	34.0	30.5	
8	5–10	38.4	34.9	26.7	35.9	10
	15–20	45.7	32.4	21.9	47.9	
9	5–10	7.5	42.5	50.0	1.3	1
	10–15	7.5	40.5	52.0	4.1	
10	5–10	9.2	35.1	55.7	0.7	0
	15–20	9.5	32.4	58.1	2.2	
11	5–10	8.8	38.0	53.2	0.5	0
	15–20	8.7	36.9	54.4	1.9	

Clay, silt and sand = % from the <2 mm fraction
EC = mmho/cm
stoniness = % from the entire soil profile

Tab. 2: *North-facing slope: particle size composition, electrical conductivity and stoniness.*

slope section whereas stone-free profiles are found at the lower part of the slopes where loess tends to accumulate.

Similar trends, although less pronounced, are displayed by the south-facing slope (tab.3).

5 Analysis

5.1 Relationships among floristic and abiotic variables

A representation of the spatial variation in the values of environmental parameters within the PCA space (figs.6 & 7) reveals the following major divisions in floristic variation. On the north-facing slope, the left side of the PCA space is

site	depth (cm)	clay	silt	sand	EC	Average stoniness
1	5–10	20.2	37.9	41.9	29.5	55
	15–20	32.6	33.7	43.7	13.8	
2	5–10	5.2	45.9	48.9	0.7	18
	15–20	12.4	37.0	50.6	5.9	
3	5–10	10.0	45.5	44.5	0.6	26
	15–20	19.0	46.9	34.1	1.8	
4	5–10	17.4	43.6	39.0	7.4	30
	15–20	21.2	48.4	30.4	12.5	
5	5–10	9.3	31.8	58.9	9.5	12
	15–20	15.7	28.1	56.2	17.5	
6	5–10	8.1	36.1	55.8	7.8	5
	15–20	14.9	36.4	48.7	24.9	
7	5–10	11.2	47.3	41.5	0.6	2
	15–20	17.4	45.4	37.2	1.6	
8	5–10	12.8	40.3	46.9	3.5	1
	15–20	17.4	40.9	41.7	6.2	
9	5–10	13.0	45.4	41.6	3.0	1
	15–20	16.4	40.6	43.0	15.1	

Tab. 3: *South.facing slope: particle size composition, electrical conductivity and stoniness.*

characterized by extremely low soil moisture values, whereas much higher values appear for the middle and right side of the PCA space (fig.6). Electrical conductivity values show a different pattern with high values at the central part and low values on both sides (fig.6). Combining the above results (fig.6) clarifies the relationship between soil moisture, soil salinity and floristic variation. The PCA space is subdivided into three distinct fields — one with high soil moisture and low EC values, the second with high soil moisture and high EC values, and the third with low soil moisture and high EC values. A possible interpretation of the above results is that the major axis of floristic variation follows a gradient from relatively high to extremely low soil moisture conditions. However within the high soil moisture values, salinity, as indicated by the EC values, exerts some influence on the floristic composition.

Fig.7 presents the PCA results for the south-facing slope. A good relationship can be observed among soil moisture content, stoniness and the floristic variation. These results compare well with the community properties shown in fig.3. In contrast, no clear relationship was found between the EC values and the floristic variation, therefore no isopleths were drawn.

The overall results clearly indicate that

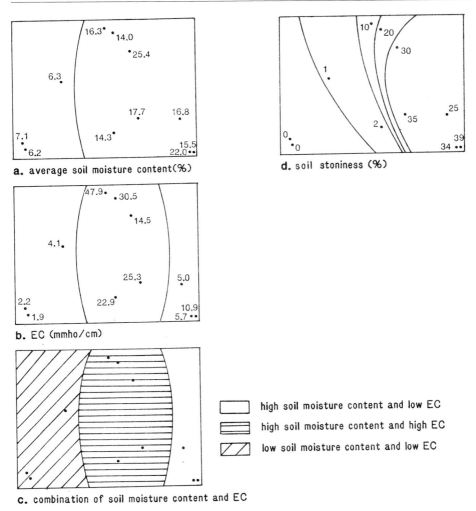

Fig. 6: *North-facing slope: relationships between PCA results and environmental factors.*

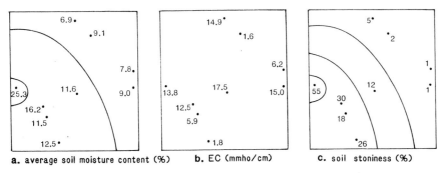

Fig. 7: *South-facing slope: relationships between PCA results and environmental factors.*

soil stoniness is the factor which can explain most of the variance in the floristic composition along the slopes. Tab.4 shows that stoniness exhibits a very high correlation coefficient also with the relative abundance of woody perennials. This fact is quite important, because the variability of woody species abundance along the slopes was the most conspicuous physiognomic phenomena.

	Species	Individuals
North-facing slope	0.96	0.87
South-facing slope	0.83	0.89
Both slopes	0.89	0.86

Tab. 4: *Correlation coefficients between stoniness and percentage of semishrub species and individuals.*

5.2 Variation of soil moisture along studied slopes

Water availability is known to be the major limiting factor in a desert environment. Spatial differences in water availability can result from the non-uniform distribution of rainfall (SHARON 1980, YAIR et al. 1980) and from the non-uniform generation of runoff which can result in spatial redistribution of water resources. Under a given wind regime the non-uniform rainfall distribution over hillslopes is mainly attributed to differences in slope angles and aspects (SHARON 1980). As these factors are quite uniform, for each of the studied slopes one can assume that rainfall distribution along these slopes is quite uniform. Under such conditions spatial differences in soil moisture can only derive from spatial differences in infiltration and runoff generation along the slopes.

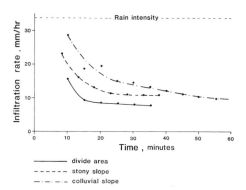

Fig. 8: *Infiltration curves of rocky and colluvial slope sections.*

5.2.1 Effects of non-uniform infiltration and runoff

Runoff plots for measuring runoff yields from the different slope sections were not installed in the frame of the present study. However, an insight into the expected differences in the frequency and magnitude of runoff generation over the rocky and colluvial slope sections was obtained through the study of the infiltration characteristics of the two basic slope units. Rainfall simulation was performed using a sprinkler developed by MORIN et al. (1970) which can accurately reproduce the characteristics of natural rainfall. Fig.8 presents the hydrological response of simulated rainfalls over the rocky and colluvial slopes. Data obtained indicate that the infiltration capacity of the rocky surface is lower than that of the soil-covered surface. As the rainstorms were simulated to produce conditions of extreme rain intensity accompanied by lengthy rain duration (KUTIEL 1978), one can expect that under natural rain conditions a more pronounced diffrence, in both infiltration and runoff generation would exist between the rocky and colluvial slope.

Infiltration data obtained can partly explain the variation in soil moisture observed along the studied slopes. The relatively quick response to rainfall over the rocky slope section results in the concentration of water from the runoff generation surfaces into the limited adjoining soil-covered areas. A consequence of water concentration is the deeper wetting of the soil, beyond that allowed by direct rainfall, creating thus an improved local water regime. Conversely, in the soil-covered areas, because of the high absorption capacity of the soil, runoff frequency and magnitude are low and any positive effect of water concentration is inhibited. The only source of water is often that of direct rainfall. As rainfall amount for most storms is quite limited, the depth of water infiltration seldom exceeds 20–40 cm and salts of airborn origin, brought by the rain, accumulate at a shallow depth (YAIR 1983). Furthermore, in the time interval that separates two consecutive storms, most of the infiltrated water is lost by evaporation and is thus of no benefit for biological activity.

5.2.2 Effect of particle size composition on soil moisture content

The water-holding capacity and water content of a soil can be greatly affected by its stoniness and clay content. Soil moisture content is usually determined for the non-gravelly fraction of the soil. Assuming a negligible water absorption capacity by the dense, low-porosity stone particles, a stony soil will display, for a given water input, a higher soil moisture content than a non-stony soil. This is due to the fact that the same water volume is absorbed by a smaller soil volume in the former than in the latter case. This explains the high correlation coefficient obtained between soil moisture and stoniness (tab.5), as well as the higher figures for moisture content over the rocky, than over the colluvial, slope section. The saturated hydraulic conductivity of stony soils can be expected to be higher than that of stoneless soils, permitting deeper infiltration in the former than in the latter, and thus providing another positive effect of stoniness upon the water regime.

	St	St + tc
North-facing slope	0.81	0.86
South-facing slope	0.96	0.98
Both slopes	0.86	0.90

| St = stoniness |
| tc = clay content |

Tab. 5: *Correlation coefficients between edaphic factors and soil moisture.*

When the clay content of the soil is taken into consideration, a slight improvement in the correlation coefficients is obtained (tab.5), pointing to the positive effect of clay content on the water-holding capacity of the soils.

6 Discussion

The main objective of this study was to determine whether the hydrological, pedological and especially botanical findings obtained at the Sede Boker experimental site could be extrapolated to an area in the Negev desert where both climatological and geological conditions differ. The comparative section that follows below will focus on the pedological and botanical aspects.

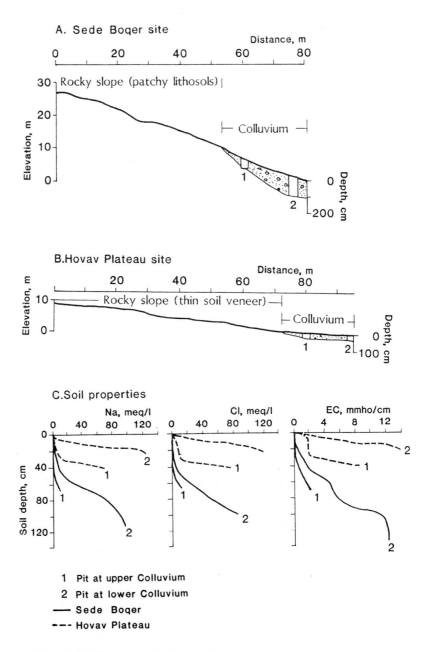

Fig. 9: *Soil properties: Hovav Plateau and Sede Boker.*

6.1 Pedological aspects

This aspect has been discussed in a recent paper (YAIR 1983). The basic information is provided in fig.9. Common to both slopes is that they face north and are each subdivided into an upper rocky and a lower colluvial section. Three factors can be expected to induce a higher amount of infiltrated water at the Hovav Plateau slope than at the Sede Boker slope. They consist of the greater annual rainfall amount (165 mm versus 92 mm, respectively), the higher sand content, and the gentler slope angle at the northern area. An increase in the amount of infiltrated water should result in better soil leaching. However, fig.9 clearly demonstrates that the southern Sede Boker soils are better leached than the northern Hovav Plateau soils. The latter are characterized by a relatively high salinity at a quite shallow depth.

Apart from the basic difference in salinity, the two soil toposequences show similar trends characterized by limited leaching and high salinity at the top of the slope, the sole source of water being direct rainfall. Leaching is more pronounced within the rocky part of the slope and upper colluvium, where water input into the soil derives from rainfall and runoff. However, leaching intensity decreases quickly downslope within the colluvial slope section where, once again, the main source of water is that of direct rainfall. It is worthwhile noting that the same pattern applies to the soil moisture regime along the slope. In the summer period, soil moisture content is always higher within the rocky areas than within the soil-covered areas (fig.5).

In summary, despite the strong similarity in the trends observed along both slopes, the northern soils located in the climatologically wetter area represents, from the pedological point of view, an environment which is far more arid than that of the climatologically drier Sede Boker area.

6.2 Botanical aspects

Tab.6 presents some vegetation characteristics that can be related to the soil/water regime. The relative importance of the perennial vegetation (expressed by cover or number of species) has a strong positive correlation with the water regime in a desert environment (SHMIDA et al. 1986). Similarly, the presence of species with Mediterranean chorotypes is indicative, in the same environemnt, of an improved water and low-salinity regime. In contrast, dominance of Saharo-Arabian species expresses real desert conditions (DANIN et al. 1975, YAIR & DANIN 1980). An analysis of tab.6 highlights the following main points:

- The number of perennial species, as well as the water requirements of the species identified, are higher over the rocky than over the colluvial slope sections.

- In both areas the lower colluvial section represents the drier environment along the slope.

- The most striking result is the difference in vegetation characteristics between the slopes. The vegetation cover, the number of woody plants and the relative importance of the Mediterranean chorotypes are much higher at the southern, climatologically drier area, and it is the northern, climatologically wetter, area which is dominated by

| | rocky section | | colluvial section | |
	Sede Boker	Hovav Plateau	Sede Boker	Hovav Plateau
C	30	10	10	0
Sp	27*	11	no data	2
Sch	18+	7	14+	0
Smd	22+	18	7+	0
Ssa	33+	55	42+	100

C = cover of perennial vegetation (%)
Sp = number of perennial species
Sch = number of semishrub species
Smd = percentage of perennial species belonging to the Mediterranean, Mediterranean-Irano-Turanian, and Mediterranean-Saharo-Arabian chorotypes
Ssa = percentage of the perennial Saharo-Arabian species
+ = After YAIR & DANIN (1980)
* = After OLSVIG-WHITTAKER et al. (1983)

Tab. 6: *Floristic parameters: a comparison between Sede Boker and Hovav Plateau.*

Saharo-Arabian species. This trend is especially pronounced over the colluvial slope sections of the two slopes.

7 Conclusions

The pedological and botanical study of the Hovav Plateau area supports the hypothesis that the findings derived from studies conducted at the Sede Boker research site can be projected to the regional scale and be applied to an area which differs in physiognomic properties. The idea that water availability, in a non-sandy desert, depends mainly on the ratio of bare-bedrock to soil cover, identified at Sede Boker, was found valid at the regional scale. This ratio is much higher in the rocky Negev Highlands than in the loess-covered plains of the Hovav Plateau. Under such conditions runoff frequency and magnitude, and therefore water availability, are higher in the southern than in the northern area, explaining thus the greater environmental aridity of the climatologically wetter Hovav Plateau.

The same principle — the ratio of bare-bedrock to soil cover — explains the variation of soil and vegetation properties along the slope. For both slopes studied, the rocky slope section, is characterized by good leaching of the soils with plants benefitting from a relatively improved water regime. The converse applies to the soil-covered slope section, where the ratio is low to very low. In this section, leaching intensity, soil moisture and vegetal cover decrease quickly in the downslope direction parallel to the increase in soil cover.

Acknowledgements

Grateful thanks are due to Mrs. T. Sofer and Mrs. M. Kidron of the Department of Geography of the Hebrew University of Jerusalem for drawing the illustrations.

References

COTTAM, G., GOFF, F.G. & WHITTAKER, R.H. (1978): Wisconsin Comparative Ordination. In: Ordination of Plant Communities. (Ed. R.H. WHITTAKER), The Hague: Junk, 185–213.

COTTLE, H. (1932): Vegetation on north and south slope mountains in southwestern Texas. Ecology, 13, 121–134.

DANIN, A., ORSHAN, G. & ZOHARY, M. (1975): The Vegetation of the Northern Negev and the Judean Desert of Israel. Israel Journal of Botany, 24, 118–172.

HILLEL, B. & TADMOR, N.H. (1962): Water regime and vegetation in the Central Negev Highlands. Ecology, 43, 33–44.

KUTIEL, H. (1978): The distribution of rain intensities in Israel. M.Sc. Thesis. Hebrew University of Jerusalem (in Hebrew).

MORIN, G., CLUFF, B.G. & POWERS, W.R. (1980): Realistic rainfall simulation for field investigations. Paper presented in 1978, at the 51st annual meeting, American Geographical Union, Washington, D.C.

NIE, N.H., HULL, C.H., JENKINS, J.G., STREINBRENNER, K. & BENT, D.H. (1975): Statistical Package for the Social Sciences. McGraw-Hill.

OLSWIG-WHITTAKER, L., SHACHAK, M. & YAIR, A. (1983): Vegetation patterns related to environmental factors in a Negev Desert watershed. Vegetation 54, 152–165.

ORSHAN, G. (1986): The deserts of the Middle East. In: Ecosystems of the World 12B, Hot Deserts and Arid Shrublands. Elsevier, Amsterdam, 1–28.

SHARON, D. (1980): The distribution of hydrologically effective rainfall incident on sloping ground. Journal of Hydrology, 46, 165–188.

SHMIDA, A., EVENARI, M. & NOY-MEIR, I. (1986): Hot deserts ecosystems: an integrated view. In: Ecosystems of the World 12B, Hot Deserts and Arid Shrublands. Elsevier, Amsterdam, 379–387.

TADMOR, N.H., ORSHAN, G. & RAWITZ, E. (1962): Habitat analysis in the Negev Desert of Israel. Bulletin of the Research Council of Israel, 11, 143–173.

YAIR, A. (1983): Hillslope hydrology, water harvesting and areal distribution of some ancient agricultural systems in the northern Negev Desert. Journal of Arid Environments, 6, 281–301.

YAIR, A. (1987): The environmental efects of loess penetration into the Northern Negev. Journal of Arid Environments, 13, 9–24.

YAIR, A. & DANIN, A. (1980): Spatial variation in vegetation as related to the soil moisture regime over an arid limestone hillside, Northern Negev. Israel. Oecologia (Berlin), 47, 83–88.

YAIR, A. & LAVEE, H. (1985): Runoff generation in arid and semi-arid zones. In: Hydrological Forecasting (Eds. M.G. ANDERSON & T.P. BURT). John Wiley & Sons, 183–220.

YAIR, A., SHARON, D. & LAVEE, H. (1980): Trends in runoff and erosion processes over an arid limestone hillside, Northern Negev, Israel. Hydrological Science Bulletin, 25, 243–255.

ZANGVIL, A. & DRUIAN, P. (1983): Meteorological data for Sede Boker. Blaustein Institute for Desert Research. Desert Meteorology Paper, Series A. No. 8

Addresses of authors:
R. Kadmon & A. Danin
Department of Botany
Institute of Life Sciences
The Hebrew University
Jerusalem, Israel 91904
A. Yair
Department of Physical Geography
Institute of Earth Sciences
The Hebrew University
Jerusalem, Israel 91904

Jan de Ploey (Ed.)

RAINFALL SIMULATION, RUNOFF and SOIL EROSION

CATENA SUPPLEMENT 4, 1983

Price: DM 120,–/US $ 75,–

ISSN 0722–0723 ISBN 3–923381–03–4

This CATENA–Supplement may be an illustration of present-day efforts made by geomorphologists to promote soil erosion studies by refined methods and new conceptual approaches. On one side it is clear that we still need much more information about erosion systems which are characteristic for specific geographical areas and ecological units. With respect to this objective the reader will find in this volume an important contribution to the knowledge of active soil erosion, especially in typical sites in the Mediterranean belt, where soil degradation is very acute. On the other hand a set of papers is presented which enlighten the important role of laboratory research in the fundamental parametric investigation of processes, i.e. erosion by rain. This is in line with the progressing integration of field and laboratory studies, which is stimulated by more frequent feed-back operations. Finally we want to draw attention to the work of a restricted number of authors who are engaged in the difficult elaboration of pure theoretical models which may pollinate empirical research, by providing new concepts to be tested. Therefore, the fairly extensive publication of two papers by CULLING on soil creep mechanisms, whereby the basic force-resistance problem of erosion is discussed at the level of the individual particles.

All the other contributions are focused mainly on the processes of erosion by rain. The use of rainfall simulators is very common nowadays. But investigators are not always able to produce full fall velocity of waterdrops. EPEMA & RIEZEBOS give complementary information on the erosivity of simulators with restricted fall heights. MOEYERSONS discusses splash erosion under oblique rain, produced with his newly-built S.T.O.R.M–1 simulator. This important contribution may stimulate further investigations on the nearly unknown effects of oblique rain. BRYAN & DE PLOEY examined the comparability of erodibility measurements in two laboratories with different experimental set-ups. They obtained a similar gross ranking of Canadian and Belgian topsoils.

Both saturation overland flow and subsurface flow are important runoff sources and the rainforests of northeastern Queensland. Interesting, there, is the correlation between s colour and hydraulic conductivity observed by BONELL, GILMOUR & CASSELLS. Rune generation was also a main topic of IMESON's research in northern Morocco, stressing t mechanisms of surface crusting on clayish topsoils.

For southeastern Spain THORNES & GILMAN discuss the applicability of erosi models based on fairly simple equations of the "Musgrave-type" After Richter (Germar and Vogt (France) it is TROPEANO who completes the image of erosion hazards in Europe vineyards. He shows that denudation is at the minimum in old vineyards, cultivated w manual tools only. Also in Italy VAN ASCH collected important data about splash erosion a rainwash on Calabrian soils. He points out a fundamental distinction between transpc limited and detachment-limited erosion rates on cultivated fields and fallow land. Fo representative first order catchment in Central–Java VAN DER LINDEN comments o trasting denudation rates derived from erosion plot data and river load measurements. H too, on some slopes, detachment-limited erosion seems to occur.

The effects of oblique rain, time-dependent phenomena such as crusting and run generation, detachment-limited and transport-limited erosion including colluvial depositi are all aspects of single rainstorms and short rainy periods for which particular, predic models have to be built. Moreover, it is argued that flume experiments may be an econo way to establish gross erodibility classifications. The present volume may give an impetu further investigations and to the evaluation of the proposed conclusions and suggestions

Jan de Ploey

G.F. EPEMA & H.Th. RIEZEBOS
  FALL VELOCITY OF WATERDROPS AT DIFFERENT HEIGHTS AS A FACTOR INFLUENCING EROSIVITY OF SIMULATED RAIN

J. MOEYERSONS
  MEASUREMENTS OF SPLASH–SALTATION FLUXES UNDER OBLIQUE RAIN

R.B. BRYAN & J. DE PLOEY
  COMPARABILITY OF SOIL EROSION MEASUREMENTS WITH DIFFERENT LABORATORY RAINFALL SIMULATORS

M. BONELL, D.A. GILMOUR & D.S. CASSELLS
  A PRELIMINARY SURVEY OF THE HYDRAULIC PROPERTIES OF RAINFOREST SOILS IN TROPICAL NORTH–EAST QUEENSLAND AND THEIR IMPLICATIONS FOR THE RUNOFF PROCESS

A.C. IMESON
  STUDIES OF EROSION THRESHOLDS IN SEMI–ARID AREAS. FIELD MEASUREMENTS OF SOIL LOSS AND INFILTRATION IN NORTHERN MOROCCO

J.B. THORNES & A. GILMAN
  POTENTIAL AND ACTUAL EROSION AROUND ARCHAEOLOGICAL SITES IN SOUTH EAST SPAIN

D. TROPEANO
  SOIL EROSION ON VINEYARDS IN THE TERTIARY PIEDMONTESE BASIN (NORTHWESTERN ITALY): STUDIES ON EXPERIMENTAL AREAS

TH.W.J. VAN ASCH
  WATER EROSION ON SLOPES IN SOME LAND UNITS IN A MEDITERRANEAN AREA

P. VAN DER LINDEN
  SOIL EROSION IN CENTRAL–JAVA (INDONESIA). A COMPARATIVE STUDY OF EROSION RATES OBTAINED BY EROSION PLOTS AND CATCHMENT DISCHARGES

W.E.H. CULLING
  SLOW PARTICULARATE FLOW IN CONDENSED MEDIA AS AN ESCAPE MECHANISM: I. MEAN TRANSLATION DISTANCE

W.E.H. CULLING
  RATE PROCESS THEORY OF GEOMORPHIC SOIL CREEP

PIPING EROSION AND BADLAND DEVELOPMENT IN SOUTH-EAST SPAIN

F. **López-Bermúdez** & M.A. **Romero-Díaz**, Murcia

Summary

Soil erosion by piping is important in the development of badlands in the semi-arid sedimentary basins of south-east Spain. This paper outlines the geological characteristics of the area, the bioclimatic conditions, badlands evolution and the infiltrability of the Neogene marl regoliths, as well as correlations between rainfall, overland flow and soil erosion. Finally, it considers the mineralogy of the illite and smectite-rich marls by x-ray analysis of samples from differents depths, in order to determine the degree of compaction and porosity of the marls. These factors control the hydraulic and hydrological characteristics of the regolith. Piping erosion is shown to be a basic process in the formation of badlands in many parts of the Mediterranean morphoclimatic domain of south-east Spain.

1 Introduction

In the arid and semi-arid environments of south-east Spain, water is the most effective agent of erosion and sediment transport. Without vegetation, all the runoff energy is directed to soil erosion and to the removal of the detached material over various distances. Aerial photographs and the landscape of this region both illustrate the most important geomorphological characteristics: the spectacular topography and its extent in badland areas, which are characterized by high drainage densities, steep slopes, low vegetation and ultimately by serious soil erosion induced by some basic denudation processes and inadequate land-use practices.

The landscape of south-east Spain, together with near-arid bioclimatic conditions, constitute a fragile environment where erosional processes are rapid and the destruction of the soils are very intense. The piping is evident in the origin and development of many gully systems and badland areas. These processes have occupied a basic role, especially where sedimentary materials are unresistant and poorly consolidated, where the soil are very porous, where the vegetation cover is scarce or absent and where man has upset the natural ecosystem. The resultant forms are widespread and found in a wide range of evolution in the sedimentary basins. Such environments are highly susceptible to this type of process.

2 Background

The importance of piping in the origin of badlands has been underlined by numerous investigators. In other countries over the last 20 years the hydrogeomorphological function of pipes has been considered as an important element in the establishment and dynamics of drainage networks and badland areas in both temperate humid and semi-arid regions (PARKER 1965, KNAPP 1970, MENSUA & IBAÑEZ 1975, STOCKING 1976, BARENDREGT & ONGLEY 1977, BRYAN et al. 1978, GALLART 1979, MORGAN 1979, GILMAN & NEWSON 1980, KIRKBY & MORGAN 1980, JONES 1981). During this period considerable advances have been made in different morphoclimatic environments, following field and laboratory observations and experiments (ROSEWELL 1970, HEEDE 1971, JONES 1971, CROUCH 1976, SHERARD et al. 1976, ZUIDAM 1976, EMBLETON & THORNES 1979, THORNES 1980, YAIR et al. 1980, RODRIGUEZ-VIDAL 1983, GUTIERREZ-ELORZA & RODRIGUEZ-VIDAL 1984, SMART & WILSON 1984, WILSON & SMART 1984, LASANTA-MARTINEZ 1985, GARCIA RUIZ et al. 1986, GERITS 1986, GARCIA-PRIETO 1986, IMESON 1986, BRYAN 1987, GOVERS 1987). Today, the problem has come a long way since CARSON & KIRKBY (1972) reported that "piping is very important for subsurface flow debris removal, but its quantitative contribution cannot yet be estimated". In south-east Spain the importance of piping in the morphology of badlands has been indicated by CANO-GARCIA (1975) in Guadix (Granada), HARVEY (1982) near Benidorm (Alicante), Sucina (Murcia), Tabernas and Vera (Almería); ROMERO-DIAZ & LOPEZ-BERMUDEZ (1985) and LOPEZ-BERMUDEZ & TORCAL-SAINZ (1986) in the areas of Mazarrón and Mula (Murcia).

These studies demonstrate that pipes can greatly influence the morphology of certain types of gullies where the fragile natural equilibrium has been disturbed by inadequate soil uses. Rapid water erosion in these areas produces subsurface tunnels, collapse features and dissection which run into the badlands and produce high rates of erosion, transport and sedimentation.

3 Description of the study area

South-east Spain, the driest part of Europe, is a geomorphological and bioclimatic region contained by the 350 mm isohyet. It has an area of some 17,200 km^2 (fig.1) with badlands occupying approximately 3,200 km^2.

3.1 The geological context of badland areas

The materials of the badlands are basically lithologies of low erosional resistance which fill the intramontain basins to great depths, unconsolidated Quaternary deposits and weakly developed soils with a poor structure. The thin irridescent gypsiferous marls, which occupy large areas, are particularly sensitive to the processes of water erosion which produce the high rates of erosion and the very dissected topography. The Cretaceous marl-sandstone, marl, and sandstone facies, and the Neogene-Quaternary materials are patchy but appear over broad surfaces. The main development in depth and extent was reached in the interior basins which were

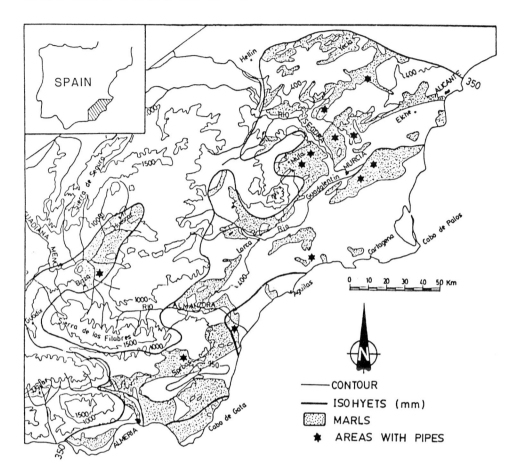

Fig. 1: *Areas of marls in south-east Spain and localities with strong piping activity.*

areas of strong subsidence and were separated following the main folding of the Betic Cordilleras. In some of these depressions, such as the Guadix-Baza-Huéscar basin, the sedimentary fill exceeds 3,000 m (VERA 1972).

Overall in these areas, badlands developed having undergone rapid evolution by erosion processes, particularly by piping, in those places where materials had high infiltration capacities and marked hydraulic gradients.

3.2 The bioclimatic conditions

South-east Spain is a Mediterranean region marginal to both the temperate climatic zone of the mid-latitudes and the subtropical desert zone. Here, the most evident climatic characteristics are the low mean annual rainfall (less than 350 mm) and an annual summer drought.

As with the whole Mediterranean domain, the observed repetition of intense and violent storms are the root of the unquestionable effectiveness of mechanical geomorphic processes (ROSSELLO-

VERGER 1986). Some 80% of the annual rainfall occurs in autumn and spring, with the greatest concentration in October. The interannual variability is about 35%. Spatial variability is extreme and corresponds to the sporadic and intense rainfall of thunderstorms, with intensities of 200 mm or more in 24 hours. The geomorphic impact of these storms on soils on unvegetated steep slopes is enormous.

The climate is semi-arid and arid in some areas. The mean annual temperatures range between 17 to 20°C and the annual potential evapotranspiration varies from 800 to 1150 mm. The hydrological deficit is very marked and affects the character of the vegetation structure. Sclerophyllous vegetation has a marked role by being the best adaptor to the low rainfall and summer drought. The vegetation communities which colonize the soils are patchy, their physiognomy corresponds to an open matorral with xerophytic herbaceous annuals and perennials, leaving large open spaces of bare ground.

The **Resogols** (FAO-UNESCO 1974), **Serozems** (KUBIENA 1952) or **Torriorthents Xerics** (SOIL SURVEY STAFF 1975), are the most important soils in badland areas, being formed from unconsolidated and poorly developed materials. They have a simple profile with the form A/C or A/AB/C, a subangular or polyhedric structure, low contents of organic material and are light in colour.

These clay-marl soils have an important soluble salt content which provides a high base exchange capacity, especially for calcium carbonate (60%) and sodium 40–70% (meq l^{-1}). Water percolation and the corresponding leaching of salts leads to dispersion of the materials and its easy entrainment and transport by subsurface flow. The pH of soil throughout the profile varies between 7.7 and 8.5. The soils are moderately plastic when wet, friable when damp and loose when dry (ROSA et al. 1984, ALIAS-PEREZ et al. 1986).

4 The badlands: a form of eroded Mediterranean landscape

In the badland areas of south-east Spain runoff is seasonal and ephemeral, following the sporadic rainfall in September, October, November and April. The development of the badlands varies in character and in intensity according to the Mediterranean environmental conditions of strong seasonal contrasts specified in each area. Their maximum extent is reached in the Tabernas area, the Vera basin, the Guadix-Baza basin, the Mula-Fortuna basin and the hydrological basins of some major rambla tributaries of the Guadalentin and Segura rivers, where weak lithology, high rainfall and inappropriate land use (deforestation, overgrazing, ploughing, etc.) favour this type of denudation.

4.1 Antiquity of the badlands

The badlands are mainly developed on marine Neogene deposits, and on those areas where the detritic Quaternary lithologies have been removed or eroded by fluvial dissection. Some areas of badlands are relatively young, recent or actively forming in the geological time scale, whilst others appear to be very old.

Around the Spanish riverine areas of the Mediterranean, the Würm glaciation (or **pluvial** in these latidudes) was a period of more dynamic geomorphological conditions than at present, which would have led to active erosion and sedimentation. The most marked pluvial phase occurred between 11,000 to 7,000 BC with sporadic incursions of summer rainfall and monsoons as far as the northern Mediterranean (FAIRBRIDGE 1968).

According to BUTZER (1963) the maximum denudation of the badlands must be attributed to rainfall increases in the past. Sedimentary evidence exists of an increase of rainfall in Mediterranean regions, which must have provoked significant water erosion such as the transport of mechanically-detached debris.

VITA-FINZI (1969), in his important study of circum-Mediterranean alluvial chronology, including examples from south-east Spain, points out the existence in some areas of two depositional units, the "Older Fill" and the "Younger Fill". The first was a period of channel erosion dated from approximately 8,000 BC to post-Roman times. More recent evidence, obtained by WISE et al. (1982) and THORNES & GILMAN (1983) suggest that the badlands around archeological sites in south-east Spain, were formed by about 2,000 BC but this does not necessarily exclude other models of geomorphological activity.

4.2 Recent Evolution: a more anthropogenic than climatic change

Geomorphological processes in the Mediterranean world since the Holocene have an important anthropogenic component due to the long human occupation of these areas (ROSSELLO-VERGER 1986). As the Mediterranean region belongs to those regions of the oldest cultivation of the world, the zonal vegetation yielded to cultivation, and forested slopes were felled and used for grazing. Both activities led to high soil erosion even though, on occasion, anthropogenic activity ceased or suppressed erosion by gullies and ramblas (riverbeds) because of the construction of dry stone walls. Examples of such areas include south-east Spain, the Balearic Islands and the Alpujarras (Andalucia).

VITA-FINZI's (1969) study of the erosion deposition chronology shows that, except for a phase of moderate deposition in the medieval period between approximately 700 and 1300 AD., net erosion has predominated since the last glaciation (**pluvial**) and has increased from the fourteenth century to the present day. In Mediterranean Spain, the great agricultural expansion of the eighteenth century and the strong rural exodus in the 1960s and 1970s, which led to the abandonment of cultivated fields in dry land farming (**secano**) and the traditional uses of ephemeral runoff, favoured the processes of water erosion, particularly of piping. At present these processes are very active and predominate in many **secano** cultivation areas abandoned 10 to 15 years ago.

5 Variations in infiltration, surface runoff and erosion in badland areas with piping

5.1 Infiltration values

The infiltration capacity in marl soils, where pipes are found, determine the quantity of subsurface water available for the formation and evolution of the pipes. These soils in general, have a lower infiltration capacity than other soil types as shown by GILMAN & THORNES (1985). The marls have low storage volumes compared to other rock types due to a fine texture and low porosity. Nev-

Soil Type: Regolith Material	Storage Volume (cm)	Infiltration Rate After 1 Hour (cm min^{-1})
Weathered surface marls	1.98 to 4.74	0.10 to 0.29
Unweathered subsurface marls	0.60 to 4.60	0.04 to 0.12
Stone-covered marls	2.04 to 8.45	0.24 to 0.42
Vegetated marls	0.00 to 3.54	0.23 to 1.49

Tab. 1: *Infiltration variability on marls.*

Source	Area	Infiltration Rate cm h^{-1}
THORNES (1976)	Various	0.30 - 1.62
SCOGINGS (1982)	Ugijar (Granada)	2.94 - 18.06 $\bar{x} = 10.08$
HARVEY (1982)	Vera-Tabernas (Almería)	2.40 - 22.20
FRANCIS (1985)	Ugijar (Granada)	3.50 - 56.07 $\bar{x} = 28.30$
LOPEZ-BERMUDEZ et al. (1985)	Mula basin (Murcia)	$\bar{x} = 8.33$
GILMAN & THORNES (1985)	Almería - Granada	$\bar{x} = 2.85$
BRU-RONDA & CUENCA-PAYA (1986)	Alicante	$\bar{x} = 6.90$
ROMERO-DIAZ (1986)	Huescar-Cullar (Granada)	7.81 - 11.05
FRANCIS (1986a)	Cañada Honda (Murcia)	7.50 - 19.96

Tab. 2: *Infiltration values in regolith soil.*

ertheless all the lithologies have high coefficients of variation which indicate that within each rock type there is a wide range of storage conditions. Infiltration variability on marls was demonstrated by LOPEZ-BERMUDEZ et al. (1985) for a series of measurements on the same Neogene marl under different conditions (tab.1).

The results show that on vegetated marls the infiltration rates are relatively hig despite a low storage capacity whilst where the infiltration rate is low the storage volumes appear greater. Various investigations have undertaken infiltration experiments in south-east Spain in order to understand their hydrological responses. Tab.2 compares results obtained using a ring infiltrometer (HILLS 1970).

The results are highly variable, ranging from 0.30 to 56.07 cm h^{-1}. Nevertheless the most significant and valid infiltration rates for piped areas are those obtained by LOPEZ-BERMUDEZ et al. (1985), GILMAN & THORNES (1985) and FRANCIS (1986b) due to the number of observations. The mean value is estimated to be between 5 and 10 cm h^{-1}.

5.2 Correlations between rainfall, surface runoff and soil loss

The relationships between rainfall, surface runoff and sediment production are complex and reflect the spatial and temporal variability of rainfall and the different soil surface conditions.

Observations and measurements in the Mula basin, undertaken over a three year period, on an experimental plot of

YEAR	J	F	M	A	M	J	J	A	S	O	N	D	Year
1984													
Rainfall (mm)	—	4.7	12.8	40.8	35.9	6.4	—	—	—	11.4	30.1	—	142.1
Sediment production (g/m²)	—	0.05	0.01	2.17	4.05	—	—	—	—	0.003	1.93	—	8.24
Runoff (mm/m²)	—	0.01	0.002	0.46	0.12	—	—	—	—	0.003	0.31	—	0.90
1985													
Rainfall (mm)	5.3	85.3	6.0	55.7	—	—	—	—	17.5	31.3	58.5	22.7	282.3
Sediment production (g/m²)	—	25.79	0.002	66.9	—	—	—	—	40.7	139.2	7.64	2.73	283.0
Runoff (mm/m²)	—	1.07	0.002	0.9	—	—	—	—	0.41	0.49	0.81	0.28	3.95
1986													
Rainfall (mm)	9.4	1.8	43.9	9.5	55.5	—	10.9	—	60.8	235.9	6.3	3.6	437.6
Sediment production (g/m²)	3.15	—	3.19	0.58	7.32	—	4.96	—	36.49	172.89	—	—	228.6
Runoff (mm/m²)	0.46	—	0.46	0.05	0.46	—	0.28	—	0.46	1.26	—	—	3.44

Tab. 3: *Rainfall, sediment production and runoff in an experimental site (Murcia, Spain).*

3,000 m² (mean slope of 28%), show a general correspondence between the three variables, although with some very significant exceptions. Tab.3 and fig.2 give the values and correlations for the three parameters. The data show that the variability of rainfall is very high; in 1984 the annual rainfall did not reach 150 mm, whilst in 1986 it was 438 mm. This is typical of an extreme Mediterranean climate. The monthly values of higher runoff, in general, coincide with the months of greater rainfall, although there are some exceptions; the degree of correlations is low, and there are notable variations between years. These points show that rainfall is not the sole control of runoff.

The above is also appricable to sediment production. Generally, the greatest soil losses occur during the months with the highest rainfall, especially those with high hourly intensities, such as the month of October. Nevertheless in months having had significant rainfall, such as April and November 1984, November and December 1985, and January and April 1986, there was barely any sediment production. Such observations confirm that the degree of correlation between rainfall, runoff and sediment production is low in semi-arid south-east Spain. Other controls are still only partly understood (FISCHER et al. 1985) such as annual variations in the content and retention of soil moisture, biomass production, the incorporation of organic material, the quantities of soluble anions and cations, conductivity, etc., and must play an important role in the complex relations between rainfall, runoff and sediment production.

In terraces of **secano** agriculture, abandoned about 10 years ago, pipes are

Photo 1: *Erosional processes in piping.*
Above: Collapse of the roof by growth of pipes and superficial cavites.
Below: Resurgence of a pipe, at the foot of a scarp between an upper and lower abandoned terrace.

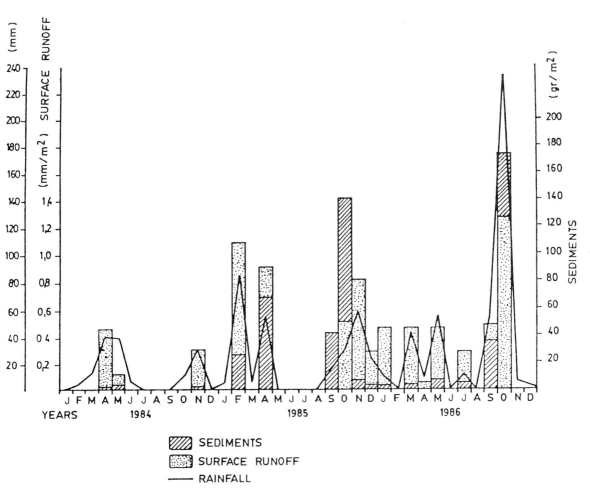

Fig. 2: *Relation between rainfall, overland flow and sediments production an experimantal plot. Mula basin, Murcia.*

actively developing (photo 1). A plot 40 × 15 × 6 m has been excavated in the central part, that is 16% of its volume involving a mean rate of soil loss and sediment production of 22.5 m³ yr⁻¹ (equivalent to a mean rate of incision of 37.5 mm yr⁻¹). This rate exceeds those obtained in other areas, such as 17.9 mm yr⁻¹ measured in the South Dakota badlands (SCHUMM 1956) or 17.3 mm yr⁻¹ from badlands in Hong Kong (LAM 1977). One must bear in mind that the value obtained in the Mula basin corresponds to a short time period and to an area particularly sensitive to the processes of water erosion. This is because they are abandoned cultivated terraces having high permeability in its non-cohesive materials and having a high content of soluble salts. The values cannot be generalized to cover wide geographical areas nor, of course, even to south.east Spain, as they are valid only for a very local area.

6 Mineralogy of the soil Regolith with piping

In order to improve our knowledge of the materials in which piping developed, X-ray diffraction techniques were applied to the mineralogy of the samples using a Philips diffractometer model PW1050/81 (Cu Kα radiation), and a Scanning Electron Microscope (SEM). The SEM used was a IST model MSM5 mini-SEN. Three samples were taken in the vertical profile of the pipes; surface, and 2 and 4 m depths. The samples were from marls with sands, containing calcite and dolomite carbonates in the form of bioclasts in the order of 50–60% by weight. They also contain small quantities of quartz and a few feldspars (fig.3). These proportions are almost constant throughout the profile with the exception of quartz, which increases slightly with depth.

The clay fraction (oriented aggregates) is scarce and badly crystallized and increase with depth. The clay consists mainly of illite and smectite aggregates. With depth, the proportion of the former decreases, as the quantity of the latter increases. Given this, the detention, or at least the slowing down of the linear incision observed in the pipes at given depths, and the activation of the processes of lateral erosion and the consequent broadening of the pipes is, among other factors, due to the increase of quartz and smectite content with depth. Using a SEM, some traits of considerable interest are apparent from the study of the samples (photo 2). At the surface the soil materials are considerably altered by dissolution, as would be expected, frequently forming aggregates separated by large, irregularly distributed pores. The carbonates are generally highly-fractured bioclasts (coccoliths). At 2 m depth, the sample appears to be better conserved as the bioclasts are easily noticed. Many pores of different sizes and irregularly distributed are still present, although fewer in number than at the surface. At 4 m depth, the soil is much more compact than in the two previous cases; practically the entire sample is composed of bioclasts, the grade of cementation is greater and its porosity is lower as compared to the upper levels.

The main change in the profile is the decrease in the number and size of the pores with depth, accompanied by better conservation of the coccoliths. Consequently, water movement becomes ever more difficult with depth and ultimately the decrease in porosity and permeability provides an impermeable horizon allowing moisture to accumulate. This in turn encourages the enlargement of pipes above the impermeable horizon.

7 Discussion and conclusions

Piping erosion appears to occupy an important role in the formation of certain types of badlands in different areas in south-east Spain. Here, piping or tunnel erosion, and some mass movements combined with fluvial processes, lead to the development of very dissected topography and high erosion rates. The initiation and development of badlands, in severe bioclimatic conditions such as found in this Mediterranean region, is fairly old, and increased following the Würm during episodes of greater mor-

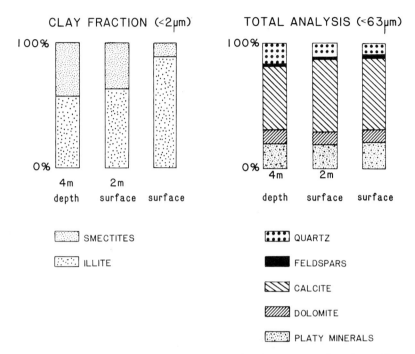

Fig. 3: *Clay mineralogy at three different levels, in a marl soil profile with active piping.*

phogenic activity. This region, occupied by man from an early stage, favoured the development of badlands as a result of changes in soil use which led to alterations in soil structure and hydraulic behaviour. The depopulation of rural areas, and the abandonment of **secano** fields of cultivation, have accelerated the process of water erosion, particularly by piping.

The variability of infiltration on the marls, where piping and badlands occur, is high as a result of the physical and ecological characteristics of the soil, and ranged from 2.85 to 19.96 cm h^{-1}. The relationships between rainfall, runoff, and sediment production during three consecutive years of monitoring appear complex and show only a general correlation, with frequent and significant exceptions. Numerous examples indicate that rainfal is not the only control affecting runoff and sediment production.

In one area of piping, situated on terraces cultivated until 1976 and since abandoned, high rates of soil loss have been obtained, of the order of 37.5 mm yr^{-1}, the highest known at present. However the results cannot be extrapolated over larger areas.

Finally, the mineralogy of the marls in which these processes occur includes a high carbonate content, and a low clay fraction constituted basically of illite and smectite whose proportions vary with depth. They are also poorly crystallized. At the surface, the soil is greatly altered with a large number of macropores facil-

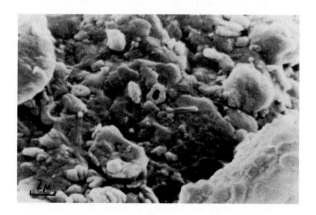

Photo 2: *SEM photographs of the samples for three levels in the soil profile.*
Upper: At surface.
Middle: At 2 m depth.
Lower: At 4 m depth at the bottom of contemporary pipes.

itating infiltration. With depth, the number and size of these intergranular pores diminishes whilst at the same time the amount of cementation and compaction increases. These data testify to the lack of homogeneity of the apparently impermeable materials such as the marl soil and to hydrological behaviour susceptible to piping and gully development.

Acknowledgements

The authors express their gratefull thanks to Prof. J. Thornes and Dr. C. Francis of the Department of Geography of Bristol University for their many helpful comments. This work is part of a wider investigation of changes of climatic and erosional processes in southeast Spain.

References

ALIAS-PEREZ, L.J. & ORTIZ-SILLA, R. (1986): Memoria de la XIV Reunión nacional de Suelos. Soc. Esp. de Ciencias del Suelo. CSIC Universidad de Murcia.

BARENDREGT, R.W. & ONGLEY, E.D. (1977): Piping in the Milk River Canyon, Southeastern Alberta, a contemporary dryland geomorphic process. In: Erosion and soil material transport in inland waters. Proceeding of the Paris Symposium. IAHS, **122**, 213–233.

BRU-RONDA, C. & CUENCA-PAYA, A. (1986): Primeras evaluaciones de la capacidad de infiltración en suelos de la comarca del campo de Alicante. En: Estudios sobre Geomorfología del sur de España. F. LOPEZ-BERMUDEZ & J. THORNES (Eds.). Universidad de Murcia, 27–30.

BRYAN, R.B. (Ed.) (1987): Rill Erosion Processes and Significance. CATENA SUPPLEMENT, **8**.

BRYAN, R.B., YAIR, A. & HODGES, W.K. (1978): Factors controlling the initiation of runoff and piping in Dinosaur Provincial Park badland, Alberta, Canada. Zeitschrift für Geomorphologie. Suppl. Band **29**, 151–168.

BUTZER, K.W. (1963): Climatic-geomorphologic interpretation of Pleistocene sediments in the Eurafrican Subtropics. In: African Ecology and Human Evolution. F.C. HOWELL & F. BOURLIERE (Eds.), Chicago, 1–27.

BUTZER, K.W. (1980): Holocene alluvial sequences: Problems of dating and correlation. In: Timescales in Geomorphology. R.A. CULLINGFORD, D.A. DAVIDSON & J. DENIS (Eds.). Wiley, Chichester, 130–142.

CANO-GARCIA, G.M. (1975): Un ejemplo de karst mecánico en rocas blandas: las torcas de Guadix. Estudios Geográficos núm., **138/139**, Madrid, 247–263.

CARSON, M.A. & KIRKBY, M.J. (1972): Hillslope Forms and Process. Cambridge University Press.

CROUCH, R.J. (1976): Field tunnel erosion — a review. Journal of the Soil Conservation Service, New South Wales, **32(2)**, 98–111.

EMBLETON, C. & THORNES, J.B. (1979): Sub-surface processes. In: Process in Geomorphology. C. EMBLETON & J.B. THORNES (Eds.). E. Arnold, London, 187–212.

FAIRBRIDGE, R.W. (1968): Holocene, Postglacial or Recent Epoch. In: The Encyclopedia of Geomorphology. Reinhold, New York, 525–536.

FAO-UNESCO (1974): Soil Map of the world. Vol 1 Legend. Roma.

FISHER, G.C., ROMERO-DIAZ, M.A., LOPEZ-BERMUDEZ, F., THORNES, J.B. & FRANCIS, C. (1985): Plant litter production and its effect in an eroding mediterranean ecosystem, Mula, SE Spain. Actas IX Coloquio de Geografía. Universidad de Murcia.

FRANCIS, C. (1985): The role and significance of surface and subsurface hydrology on gully head growth in South Spain. Doctoral Thesis. Bedford College. Department of Geography. University of London.

FRANCIS, C. (1986a): Soil erosion in fallow fields: an example from Murcia. Papeles de Geografía Física, **11**. Departamento de Geografía Física, Universidad de Murcia, 21–28.

FRANCIS, C. (1986b): Subsurface hydrology of gully heads (Ugijar Basin). In: Estudios sobre Geomorfología del sur de España. F. LOPEZ-BERMUDEZ & J. THORNES (Eds.), Universidad de Murcia, 67–70.

GALLART, F. (1979): Observaciones sobre la Geomorfología dinámica actual en la cuenca d'Odena (alrededores de Igualada, prov. de Barcelona). Actas III reunión Nac. del Grupo Español de Trabajo del Cuaternario, 123–134.

GARCIA-PRIETO, F.J. (1986): Dinámica erosiva del piping: Un ejemplo en la Depresión del Ebro. Cuadernos de Investigación Geográfica, T. 12, fasc. 1–2 Logroño, 11–24.

GARCIA-RUIZ, J.M., LASANTA, T., ORTIGOSA, L. & ARNAEZ, J. (1986): Pipes in cultivated soils of La Rioja: origin and evolution. Zeitschrift für Geomorphologie. Suppl. Bd. 58, 93–100.

GERITS, J.J.P. (1986): Regolith properties and badlands development. En Estudios sobre geomorfología del sur de España. F. LOPEZ-BERMUDEZ & J. THORNES (Eds.). Universidad de Murcia, 71–74.

GILMAN, K. & NEWSON, M.D. (1980): Soil pipes and pipe flow: a hydrological study in upland Wales. British Geomorphological Research Group, Monograph no. 1.

GILMAN, A. & THORNES, J.B. (1985): Land-use and Prehistory in Southeast Spain. The London research Series in Geography, 8. George Allen & Unwin, London.

GOVERS, G. (1987): Spatial and temporal variability in rill development processes at the Huldenberg experimental site. CATENA SUPPLEMENT, 8, 17–34.

GUTIERREZ-ELORZA, M. & RODRIGUEZ-VIDAL, J. (2984): Fenómenos de sufusion (piping) en la Depresión Media del Ebro. Cuadernos de Investigación Geográfica, T.10; fasc. 1-2. Logroño, 75–83.

HARVEY, A. (1982): The role of piping in the development of badlands and gully systems in south-east Spain. In: Badland geomorphology and Piping. R.B. BRYAN & A. YAIR (Eds.). Geo Books, Norwich, 317–335.

HEEDE, B.H. (1971): Characteristics and processes of soil piping in gullies. U.S. Department of Agriculture, Forest Service Research Paper. R.M. 68.

HILLS, R.C. (1970): The determination of the infiltration capacity of field soils using the cylinder infiltrometers. British Geomorphological Research Group, Technical Bulletin, 3.

IMESON, A.C. (1986): Investigating volumetric changes in clayey soils related to subsurface water movement and piping. Zeitschrift für Geomorphologie, Suppl. Bd., 59, 115–130.

JONES, J.A.A. (1971): Soil piping and stream channel initiation. Water Resources Research, 7(3), 602–610.

JONES, J.A.A. (1981): The nature of soil piping. A review of research. British Geomorphological Research Group, Research Monograph, 3, Norwich.

KIRKBY, M.J. & MORGAN, R.P.C. (1980): Soil Erosion. John Wiley and Sons, Chichester.

KNAPP, B.J. (1970): A note on throughflow and overland flow in steep mountain watersheds. Reading Geographer, 1, 40–43.

KUBIENA, W.L. (1952): Claves sistematica de suelos. C.S.I.C., Madrid.

LAM, K.C. (1977): Patterns and rates of slopewash on the badlands of Hong Kong. Earth Surface Processes, 2, 319–322.

LASANTA-MARTINEZ, T. (1985): Aportación al estudio de la erosión hídrica en campos cultivados de La Rioja. Instituto de Estudios Riojanos. Logroño.

LOPEZ-BERMUDEZ, F., ROMERO-DIAZ, M.A., RUIZ-GARCIA, A., FISHER, G.C., FRANCIS, C. & THORNES, J.B. (1985): Erosión y ecología en la España semi-árida (Cuenca de Mula, Murcia). Cuadernos de Investigación Geográfica, X. Logroño, 113–126.

LOPEZ-BERMUDEZ, F. & TORCAL-SAINZ, L. (1986): Procesos de erosión en tunel (piping) en cuencas sedimentarias de Murcia (España). Estudio preliminar mediante difracción de rayos X y microscopio electrónico de barrido. Papeles de Geografía Física, 11. Universidad de Murcia, 7–20.

MENSUA, S. & IBAÑEZ, M.J. (1975): Alveolos en la Depresión del Ebro. Cuadernos de Investigación Geográfica, T.2; 3–14.

MORGAN, R.P.C. (1979): Soil Erosion. Topics in Applied Geography. Longmans, London.

PARKER, G.C. (1965): Piping, a geomorphic agent in landforms development of the drylands. International Association Scientific Hydrology, 65, 103–113.

RODRIGUEZ-VIDAL, J. (1983): Geomorfología de las Sierras Exteriores oscenses y su piedemonte. Tesis Doctoral. Facultad de Ciencias. Universidad de Zaragoza.

ROMERO-DIAZ, M.A. (1986): Variaciones espaciales de infiltración y su relación con la textura de los suelos en al NE de la provincia de Granada. En: Estudio sobre Geomorfología del sur de España. F. LOPEZ-BERMUDEZ & J. THORNES (Eds.), Universidad de Murcia, 121–126.

ROMERO-DIAZ, M.A. & LOPEZ-BERMUDEZ, F. (1985): Procesos de erosión en la Cuenca Neógena-Cuaternaria de Mula. En Guia de Itinerários Geográficos de la región de Murcia. IX Coloquio de Geografos Españoles, Murcia, 83–98.

ROSE de la D., BAÑOS, C., MUDARRA, J.L., BARAHONA, J.M., MOREIA, J.M., GAGO, R., PUERTAS, J.M. & RAMOS, A. (1984): Catalogo de suelos de Andalucia. Agencia de Medio Ambiente. Sevilla.

ROSEWELL, C.J. (1970): Investigations into the control of earthwork tunnelling. Journal of Soil Conservation Services, New South Wales, 26(3), 188–203.

ROSELLO-VERGER, V.M. (1986): Ramblas y barrancos: un modelo de erosión Mediterránea. Ponencia al IX Coloquio de Geografos Españoles. Universidad de Murcia, 117–184.

SCHUMM, S.A. (1956): Evolution of drainage systems and slopes in badlands at Perth Amboy, New Jersey. Bulletin of Geological Society of America, 67, 597–646.

SCOGINGS, H.M. (1982): Spatial variations in infiltration, runoff and erosion on hillslopes in semi-arid Spain. In: Badland Geomorphology and Piping. R.B. BRYAN & A. YAIR (Eds.). Geo Books, Norwich, 89–112.

SHERARD, J.L., DUNNIGAN, K.P. & DECKER, R.S. (1976): Identification and nature of dispersive soils. J. Geotechn. Div. Proc. Am. Soc. Covil Engrs. 102(GT4), 287–301.

SMART, P.L. & WILSON, C.M. (1984): Two methods for the tracing of pipe flow on hillslopes. CATENA, 11, 159–168.

SOIL SURVEY STAFF (1975): Soil taxonomy: A basic system of classification for making and interpreting soil survey. Agr. Handbook No. 436. USDA. Soil Cons. Serv., Washington.

STOCKING, M.A. (1976): Tunnel erosion. Rhodesia Agricultural Journal, 73(2), 35–39.

THORNES, J.B. (1976): Semi-arid Erosional Systems: Case Studies from Spain. Geography Papers, 7, London School of Economics.

THORNES, J.B. (1980): Erosional processes of runnign water and their spatial and temporal controls: a theoretical viewpoint. In: Soil Erosion. M.J. KIRKBY & R.P.C. MORGAN (Eds.). John Wiley and Sons, London, 129–182.

THORNES, J.B. & GILMAN, A. (1983): Potential and actual erosion around archaeological sites in south east spain. In: Rainfall Simulation, Runoff and Soil Erosion. J. De PLOEY (Ed.). CATENA SUPPLEMENT, 4, 91–113.

VERA, J.A. (1972): Mapa Geológico de España E. 1/200.000. Hoja num.78 (Baza). I.G.M.E. Madrid.

VITA-FINZI, C. (1969): The Mediterranean Valleys. Geological Changes in Historical Times. Cambridge University Press.

WILSON, C.M. & SMART, P.L. (1984): Pipes and pipe flow process in an upland catchment, Wales. CATENA, 11, 145–158.

WISE, S.M., THORNES, J.B. & GILMAN, A. (1982): How old are the badlands. A case study from South-East Spain. In: Badland geomorphology and Piping. R.B. BRYAN & A. YAIR (Eds.). Geo Books. Norwich, 259–277.

YAIR, A., BRYAN, R.B., LAVEE, H. & ADAR, E. (1980): Runoff and erosion processes and rates in the Zin Valley badlands, Northern Negev, Israel. Earth Surface Processes, 5, 205–225.

ZUIDAM, R.A. (1976): Geomorphology development of the Zaragoza region, Spain. International Institute for Aerial Survey and Earth Sciences, Enschede.

Address of authors:
F. López-Bermúdez & M.A. Romero-Díaz
Departamento de Geografía Física
Universidad de Murcia
30.001 Murcia, Spain

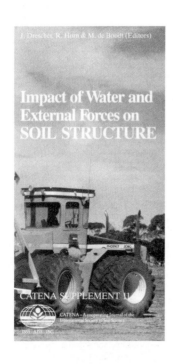

J. Drescher, R. Horn & M. de Boodt (Editors):

Impact of Water and External Forces on SOIL STRUCTURE

CATENA SUPPLEMENT 11, 1988

DM 149, — / US $88. —

ISSN 0722-0723 / ISBN 3-923381-11-5

CONTENTS

Preface

W.F. Van Impe, M. De Boodt & I. Meyus
Improving the Bearing Capacity of Top Soil Layers by Means of a Polymer Mixture Grout

H.H. Becher
Soil Erosion and Soil Structure

H.-G. Frede, B. Chen, K. Juraschek & C. Stoeck
Simulation of Gas Diffusion

H. Bohne & R. Lessing
Stability of Clay Aggregates as a Function of Water Regimes

A.R. Dexter
Strength of Soil Aggregates and of Aggregate Beds

R. Horn
Compressibility of Arable Land

K.H. Hartge
The Reference Base for Compaction State of Soils

B.G. Richards & E.L. Greacen
An Example of Numerical Modelling – Expansion of a Root Cavity in Soil

A. Ellies
Mechanical Consolidation in Volcanic Ash Soils

H.H. Becher & W. Martin
Selected Physical Properties of Three Soil Types as Affected by Land Use

I. Håkansson
A Method for Characterizing the State of Compactness of an Arable Soil

C. Sommer
Soil Compaction and Water Uptake of Plants

W. Köppel
Dynamic Impact on Soil Structure due to Traffic of Off-Road Vehicles

W.E. Larson, S.C. Gupta & J.L.B. Culley
Changes in Bulk Density and Pore Water Pressure during Soil Compression

A.L.M. van Wijk & J. Buitendijk
A Method to Predict Workability of Arable Soils and its Influence on Crop Yield

N. Burger, M. Lebert & R. Horn
Prediction of the Compressibility of Arable Land

H. Borchert
Effect of Wheeling with Heavy Machinery on Soil Physical Properties

P.H. Groenevelt
Impact of External Forces on Soil Structure

B.P. Warkentin
Summary of the Workshop

APPLICABILITY OF THE MODIFIED UNIVERSAL SOIL LOSS EQUATION IN SMALL CARPATHIAN WATERSHEDS

M. **Madeyski**, Cracow
K. **Banasik**, Warsaw

Summary

Estimates of sediment yields from watersheds are needed for reservoir design, conservation practice design and water quality prediction. Data from six small watersheds in Polish Carpathians has been used to establish the parameters α and β of the Modified Universal Soil Loss Equation. Results of the study have shown a tendency for the MUSLE to overpredict sediment yields from the investigated watersheds.

1 Introduction

Numerous field measurement of sediment yield from small watersheds have been conducted in south and southeastern Poland but no regional formula of sediment yield estimation has been worked out. Accordingly, attempts are now being made to adapt and develop foreign formulae. The application of the Universal Soil Loss Equation (USLE) (WISCHMEIER & SMITH 1978) and Sediment Delivery Ratio (SDR) (ROEHL 1962) to predict the average annual sediment yield from a small agricultural watershed in the centre of Poland has given encouraging results (BANASIK 1985). The next step in examining the applicability of the USLE is described in this paper.

The approach put forward by WILLIAMS (1975) for estimating storm event sediment yield was used in this study for the analysis of data collected from six small Carpathian watersheds (fig.1). The form of his Modified Universal Soil Loss Equation (MUSLE) is:

$$Y = \alpha (V * Q_p)^\beta * K * LS * C * P \quad (1)$$

where
Y = sediment yield from an individual storm (Mg)
α, β = coefficients (after WILLIAMS (1975), $\alpha = 11.8$ and $\beta = 0.56$)
V = storm runoff volume (m^3)
Q_p = peak runoff rate from the storm (m^3/s)
K = soil erodibility factor (USLE unit)
LS = slope length and steepness factor (1)
C = cover management factor (1)
P = erosion control practice factor (1)

The form of the MUSLE is essentially the same as the USLE except that the rainfall factor has been replaced by the runoff factor. This substitution eliminated the need for estimation of delivery ratios. The values for V and Q_p are determined from runoff data or from rainfall-runoff simulation.

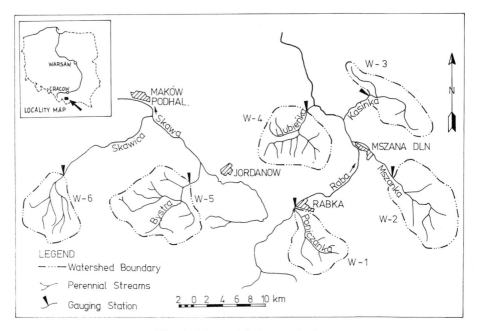

Fig. 1: *Map of the watersheds.*

The MUSLE was developed by WILLIAMS (1975) to predict sediment yields from watersheds under a variety of conditions and land uses. The equation has been included in several hydrologic and sediment yield models (SPRINGER et al. 1984, WILSON et al. 1984) and has been applied to other regions (JOHNSON et al. 1985).

The aim of this case study is the estimation of the parameters α and β for equation (1) based on the field data from the six small watersheds studied. The need for such studies has also been suggested by HADLEY et al. (1985).

2 Field data description

Data used to evaluate parameters of equation (1) were collected from the six watersheds (fig.1) by the Department of Hydraulic Structures, Cracow Agricultural University during the summer periods of 1976–1979. Some data has previously been published (MADEYSKI 1980). The values Y, V, and Q_P have been estimated on field measurements. Systematic suspended sediment sampling was carried out at the same river cross-section points, for varying time intervals (depending on discharge) from 1 to 5 hours. The suspended sediment concentration was measured simultaneously — over entire cross-sections ($C\alpha$), and at selected points (Cp) — and was used to determine the relationship $c\alpha = f(Cp)$ for each gauging station. This relationship, together with discharge Q (determined on the basis of water level records and discharge curves) permitted storm event sediment yield to be estimated. For the estimation of values V and Q_p, the base flow was not separated from the total flow.

The value of the K, LS and CP factors were estimated from soils, topographic

Water-shed	Area (km²)	Cropland ratio	USLE parameters				Number of events
			K	LS	C	P	
W-1	33.1	.40	.18	7.3	.078–.161	.65	12
W-2	51.0	.44	.17	10.8	.087–.179	.65	13
W-3	32.0	.41	.14	9.5	.079–.165	.65	13
W-4	48.7	.43	.18	10.0	.085–.174	.65	16
W-5	77.0	.26	.17	8.1	.039–.107	.65	13
W-6	48.6	.10	.12	14.0	.021–.042	.65	11

Tab. 1: *Characteristics of the watersheds.*

maps, and cropping history of the watersheds, using tables and nomographs from WISCHMEIER & SMITH (1978).

The K, LS, C and P factors were weighted according to drainage area after WILLIAMS (1975). The general form of the weighting function is:

$$X = \frac{\sum_{i=1}^{n} x_i * DA_i}{DA} \quad (2)$$

where
X = weighted factor
x_i = value of the factor covering drainage area DA_i
DA = total drainage area of the watershed

Watershed areas range from 32 to 77 km² (tab.1). Data were collected for a total of 78 individual rainfall events of depths ranging from 4.7 to 53.0 mm. The average annual rainfall is about 880 mm at W-2 and 1200 mm at W-6. The mean elevation of the watersheds ranges from 570 m at W-3, to 900 m at W-6. The two main soils are skeletal soil and loamy soil with K factor values of 0.06 and 0.21, respectively. There is significant forest cover in each watershed, ranging from 37% in W-3 to 80% in W-6. The crops cultivated include corn (oats and rye), clover and potatoes. The only erosion control practice used for some parts of the watersheds are grassed waterways.

3 Results and concluding remarks

To determine the parameters α and β of the relationship:

$$\frac{Y}{K * LS * C * P} = \alpha(V * Q_p)^\beta \quad (3)$$

the logarithmic transformation was applied with substitution:

$$Y = \log(Y/(K * LS * C * P)) \quad (4)$$

$$x = \log(V * Q_p) \quad (5)$$

$$a = \log(\alpha) \quad (6)$$

and equation (3) becomes:

$$y = a + \beta * x \quad (7)$$

The least squares method has been used to determine the values of α and β, and α was determined as numerous logarithms of a.

The determination of the parameters was carried out for two cases. For the first all 78 storm event sediment yields have been taken into consideration. For the second, only 35 events of rainfall depth not less than 20 mm have been considered. The obtained values are:

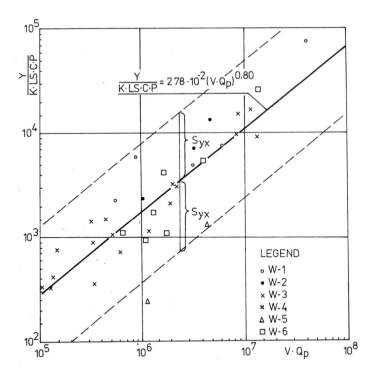

Fig. 2: *Relationship of analysed data in Case 2.*

- Case 1

$\alpha = 5.42 * 10^{-4}$
$\beta = 1.02$
r (correlation coefficient) = 0.71

- Case 2

$\alpha = 2.78 * 10^{-2}$
$\beta = 0.80$
r = 0.87

with the critical correlation coefficient, at the 5% significance level, being 0.22 and 0.33 for Case 1 and 2, respectively.

The correlation coefficient shows that a better predictive equation is obtained in Case 2. The standard error of the estimate of the regression equation (7) for Case 2 (S_{yx}) was 0.672. Data used in Case 2, the prediction relationship and the standard error of the estimate (S_{yx}) are shown in fig.2. In both cases the parameters α and β are significantly different from the ones given by WILLIAMS (1975). By comparing the prediction equations (fig.3), one can determine that the ratio of sediment yields, according to WILLIAMS (1975) and Case 2, decrease from ca. 27 for $V * Q_p = 10^5$, to ca. 5 for $V * Q_p = 10^8$.

Although the analysis has been carried out with limited data, it is useful to point out that:

- MUSLE, with parameters given by WILLIAMS (1975), overpredicts the observed results in this case study, and thus the equation should not be trans-

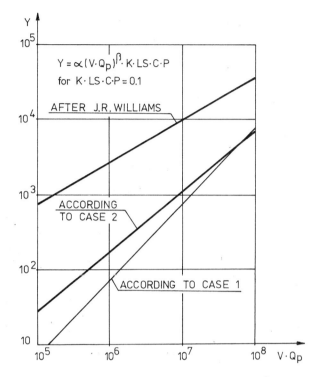

Fig. 3: *Comparison of the relationships* $Y = \alpha(V * Q_p)^\beta K * LS * C * P$.

ferred to other regions without prior verification,

- MUSLE, with parameters from Case 2, can be used to approximate sediment yield from individual storms in small watersheds of the Polish Carpathians,

- more intensive studies and new data are needed to determine more accurately the parameters α and β and to apply the equation to other regions.

References

BANASIK, K. (1985): Applicability of the universal soil loss equation for predicting sediment yield from small watersheds in Poland. Proceedings of the International Symposium on Erosion, Debris Flow and Disaster Prevention, Tsukuba, Japan, 85–88.

HADLEY, R.F., LAL, R., OSTAND, C.A., WALLING, D.E. & YAIR, A. (1985): Recent developments in Erosion and Sediment Yield Studies. Technical Documents in Hydrology, IHP-UNESCO, Paris.

JOHNSON, C.W., GORDON, N.D. & HANSON, C.L. (1985): Northwest rangeland sediment yield analysis by the MUSLE. Transactions of the ASAE **28(6)**, 1889–1895.

MADEYSKI, M. (1980): Transport of suspended load in little river basins. Proceedings of IV International Seminar on Transport and Sedimentation of Solid Particles, Wroclaw-Trzebieszowice, POland, Vol. II, ES.1–ES.8

ROEHL, J. (1962): Sediment source area, delivery ratios and influencing morphological factors. IAHS Publ. **59**, 202–213.

SPRINGER, E.P., JOHNSON, C.W., COOLEY, K.R. & ROBERTSON, D.C. (1984): Testing the SPUR hydrology component on rangeland watersheds in southwest Idaho. Transactions of the ASAE **27**, 1040–1046, 1054.

WILLIAMS, J.R. (1975): Sediment-yield prediction with universal equation using runoff energy-factor. In Present and Prospective Technology for Predicting Sediment Yields and Sources, 224–252. USDA-ARS-S-40.

WILSON, B.N., BARFIELD, B.J., MOORE, I.D. & WARNER, R.C. (1984): A hydrology and sedimentology watershed model. Part II: Sedimentology component. Transactions of the ASAE **27**, 1378–1384.

WISCHMEIER, H.W. & SMITH, D.D. (1978): Predicting rainfall erosion losses — a guide to conservation planning. USDA-ARS, Agriculture Handbook No. 573.

Addresses of authors:
M. Madeyski
Research Engineer
Cracow Agricultural University
Cracow
Poland
K. Banasik
Research Engineer
Warsaw Agricultural University
Warsaw
Poland

A MODEL FOR HEADCUT RETREAT IN RILLS AND GULLIES

J. De Ploey, Leuven

Summary

A relatively simple mathematical model has been developed to describe recession in rills and gullies. In addition to an erodibility factor, the model allows for measurable topographic and hydrodynamic parameters. It predicts the merging of small knickpoints into larger steps. The possible application of the model to large falls, such as Niagara Falls, is briefly discussed. The first calculations appear to produce acceptable values for the erodibility factor.

Resume

Un modèle mathématique relavitement simple est proposé pour la description du recul de têtes d'incision dans des rigoles et des ravins. Ce modèle introduit un facteur d'érodibilité et quelques paramètres topographiques et hydrodynamiques fondamentaux. Il prédit le passage de petits "escaliers" dans de plus grandes têtes d'incision et il semble également s'appliquer aux grandes chutes d'eau. En effet, les premiers calculs semblent produire une suite acceptable de valeurs pour le facteur d'érodibilité.

ISSN 0722-0723
ISBN 3-923381-17-4
©1989 by CATENA VERLAG,
D–3302 Cremlingen-Destedt, W. Germany
3-923381-17-4/89/5011851/US$ 2.00 + 0.25

1 Introduction

The regression of knickpoints, headcuts and falls is one of the most striking features in the development of rills, gullies and rivers. Their site generally corresponds to a marked hydraulic jump related tp plunge-pool erosion by which a vertical or subvertical face, with a height h, migrates upstream. At the base of this face a combination of vortex erosion and splash erosion provokes undercutting and subsequent soil or rock falls of cohesive material.

The commonly known large waterfalls often originated in an obscure geological past but during each rainstorm one can observe the generation of series of headcuts in rills, entrenching cohesive material. In laboratory experiments on loess loams, SAVAT (1976) has shown headcuts related to the generation of standing waves in a supercritical flow (Froude Number >1). On loamy subsoils this often seems to occur on critical slopes of 4–5% which, therefore, are very sensitive to rill erosion (SAVAT & DE PLOEY 1982). Finally, for the same silty loess loams, GOVERS (1985) made clear that the development of knickpoints and headcuts, concomitant to rill incision, occurs for critical shear flow velocities of the order of 3–4 cm/s, which means a critical shear stress of about 1 Pa.

There are reasons to believe that

plunge-pool erosion and headcut retreat are also active processes in piping which is so important in gullying and in the development of badlands (BRYAN & YAIR 1982). What is still needed however, is explicit information on its real impact.

MITCHELL & BUBENZER (1980) reviewed the data on gully advancement for the USA. Empirical equations have been proposed to predict gully growth as a function of drainage area above the gully head, rainstorm intensity, peak runoff rate, slope of the approach channel and properties of the eroding material, but there are great disparities between the proposed models. In addition mechanisms of headcut retreat are not discussed. Accordingly, the present article attempts to fill this gap. A deterministic model of headcut retreat is proposed, based on two fundamental proportionalities and the introduction of an erodibility coefficient which, of necessity, also includes a black-box field factor.

2 Model construction

In the case of rills or gullies, one may consider the effects of a single storm event in a drainage area A (fig.1). A headcut, with width w and height h, will move upstream over a distance R, with a rate $U_R = R/t$. Rainfall, with a mean intensity I_r occurs during a period t, producing a flow discharge Q corresponding to a total mass of water $M = Q t$. The flow reaches the headcut or the fall with a mean velocity $\bar{u}$ (fig.1). From fluid dynamics it is known that the speed of the water u_{pp} which strikes the plunge-pool is equal to:

$$u_{pp} = (2 g h + \bar{u}^2)^{0.5} \qquad (1)$$

The corresponding kinetic energy is:

$$M u_{pp}^2/2 = \frac{Q t(2 g h + \bar{u}^2)}{2} \qquad (2)$$

According to classic soil mechanics the stability of headcut walls or gully banks is expressed by:

$$H_c = \frac{N_S c'}{v_w} \qquad (3)$$

where
H_c = the critical height of a slope having a given slope angle;
N_S = the stability factor, a pure number whose value depends on the slope angle, on the shear strength of the material and on several structural properties;
c' = the cohesion;
v_w = the bulk density of the material.

The volume $V = R w h$, eroded by headcut advancement, will be proportional to:

$$V \sim \frac{1}{H_c} = \frac{v_w}{N_S c'} \qquad (4)$$

It may be advisable to introduce in (4) the resistance to penetration r_p, with the same dimensions as C'. Thus,

$$V \sim \frac{v_w}{N_S (c' r_p)^{0.5}} \qquad (5)$$

A cone penetrometer test will give some information on the internal stability of the material behind the banks.

Furthermore, a second proportionality is introduced:

$$V \sim \frac{M u_{pp}^2}{2} = \frac{Q t (2 g h + \bar{u}^2)}{2} \qquad (6)$$

Combining equations (4) and (6) gives:

$$V = R w h \sim \frac{Q t (2 g h + \bar{u}^2) v_w}{2 N_S c'} \qquad (7)$$

An equation is obtained after defining the erodibility factor E_r :

Model for Headcut Retreat

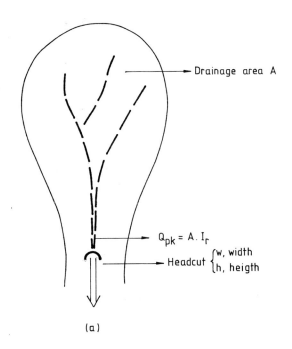

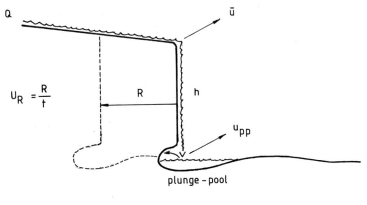

Fig. 1: *Effect of a single rainstorm event on headcut retreat.*
(a) drainage area;
(b) headcut

$$V = E_r \, Q \, t \, (2 \, g \, h + \bar{u}^2/2) \quad (8)$$

where $E_r = F v_w/c'$ or $F v_w/(c' r_p)^{0.5} F$ becomes the classic black-box field erodibility coefficient which includes the N_S value, and covers the complex interactions between the geometry of the banks and the plunge-pool as well as the mechanical and structural properties of the material.

Solving for R, equation (8) gives:

$$R = \frac{E_r \, Q \, t \, (g + \bar{u}^2/2h)}{w} \quad (9)$$

and the rate of recession is equal to:

$$U_R = \frac{E_r \, Q \, (g + \bar{u}^2/2h)}{w} \quad (10)$$

Since Q/w corresponds to the unit discharge q, we can also introduce the Reynolds Number $Re = q/n$; n being the kinematic viscosity of the water:

$$R_R = E_r \, Re \, n \, (g + \bar{u}^2/2h) \quad (11)$$

3 Data and discussion

The discussion focuses on equation (10), the rate of headcut regression in a rill or a gully, i.e. the upstream advancement of a fall on a river.

First, it must be mentioned that the impact of the second term $\bar{u}^2/2h$, in equation (10), on U_R becomes important when the ratio $\bar{u}/h$ increases. This will be the case for rills, especially rills on relatively steep slopes. Also, within the same rill, small headcuts will move upslope more rapidly than larger ones. Therefore, after a certain priod, a limited number of relatively large headcuts will result from the coalescence of many smaller ones. This evolution is commonly observed in the field and during flume experiments with silty loess loams. From these experiments, E_r values have been obtained ranging between 10^{-6} and 10^{-5} s^2/cm^2. For rills with a width of the order of 10 cm, on low to moderate slopes, U_R is often of the order of 1 cm/min. A numerical example illustrates the impact of the term $\bar{u}^2/2h$ on U_R, for a 13 cm wide and 2 cm high headcut advancing in silty loess loam during a laboratory experiment in a 1 m flume, under the following conditions: $E_R = 4 \times 10^{-6}$ s^2/cm^2; $Q = 104$ cm^3/s; $g = 981$ cm/s^2; $\bar{u} = 67$ cm/s:

$$U_R = \frac{4 \times 10^6 \times 104(981 + 1,122)}{13}$$
$$= 0.0673 cm/s$$

or 4 cm/min.

An incision of a rill on a relatively steep slope S can result from the progressive retreat and the deepening of one or more headcuts (fig.2). The slope of the rill S_c may express the mean transport capacity of the flows during competent rainstorms. Therefore the value of S_c may be determined by the colluviation model (DE PLOEY 1984). In this case

$$S_c = A \frac{c^{0.8}}{q^{0.5}} \quad (12)$$

where
A is a function of the mean grain diameter, in cm
c is the load concentration, in g/l
q is the unit discharge, in cm^2/s.

From the geometry of the system it can easily be deduced that

$$V = w \, r(h + (\tan S - \tan S_c)) \quad (13)$$

In the case of gullies with a low slope, the term $\bar{u}^2/2h$ can become negligible and is written

$$U_R = E_r \, Q \, g/w \quad (14)$$

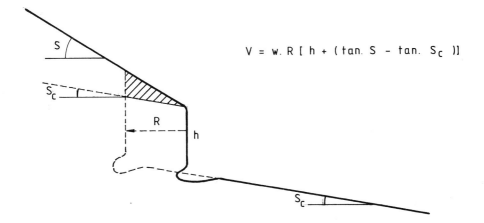

Fig. 2: *Progressive retreat and deepening of one or more headcuts.*

Few data are available in the literature about R and U_R values for large gullies. In the USA, IRELAND et al. (1939), HEEDE (1974), PIEST et al. (1975) discussed measurements of gullies, mostly developed on clayey loams, in South Carolina, Colorado and Iowa, respectively. Using their data and solving equation (14) gives E_r values of the order of 10^{-7} s^2/cm^2; one order below the silty loam of Belgium.

One may also apply equation (14) to large falls, e.g. the Niagara Horseshoe Falls where plunge-pool erosion saps shales capped by dolomites. According to CALKIN & BRETT (1978) the mean annual retreat during the last centuries has been 110 cm/year, or $U_R = 3.5 \times 10^{-6}$ cm/s.

Further data obtained are: $Q = 6.10^9$ cc/s; $w = 90,000$ cm; $h = 5700$ cm. Hence

$$E_r = \frac{U_R w}{Q g} = 5.4 \times 10^{-14} s^2/cm^2 \quad (15)$$

The Rhine Fall near Schaffhausen is entrenched on hard Jurassic limestones and the input for the calculation of E_r is: $Q = 6.10^8$ cc/s; $w = 15,000$ cm; $h = 2300$ cm; $U_R = 6.34 \times 10^{-9}$ cm/s (0.2 cm/year). According to equation (14) E_r is equal to 1.6×10^{-16} s^2/cm^2. As expected the erodibility of the Jurassic limestones is significantly lower than that of the shales of the Niagara Falls. In the case of these falls, and of large gullies, the proposed model predicts U_R values which are independent of the height h of the headcut or of the fall.

4 Conclusions

The proposed model for knickpoint and headcut retreat is relatively simple in its construction and introduces a set of basic and fairly well measurable parameters.

For large gullies and falls it predicts a linear relationship between U_R and Q (total discharge of the flow). The data of PIEST et al. (1975), on the Iowa gullies, seems to confirm this relationship but there is need for more measurements to check the model on this point. Research in the future should also focus on the nature and the analysis of the erodibility factor E_r. In this respect we should not forget to establish the link between E_r and the presence of dispersive clays in (arid) areas where piping plays an important role in gully erosion and badland evolution (IMESON & VERSTRATEN 1981, IMESON, KWAAD & VERSTRATEN 1982). According to these authors, dispersion, and hence erosion, are positively influenced by a high sodium adsorption ratio combined with a low electrolyte concentration, EC_e. Therefore, it may be necessary to investigate the relationship

$$E_r \sim SAR/EC_e \qquad (16)$$

Still, under these conditions, it is questionable in as how far E_r is positively or negatively influenced by the bulk density of the material, V_w.

References

BRYAN, R. & YAIR, A. (Eds.) (1982): Badland Geomorphology and Piping. Geo Books, Norwich.

CALKIN, P.E. & BRETT, C.E. (1978): Ancestral Niagara River drainage: stratigraphic paleontologic setting. Bulletin of the Geological Society of America, **89**, 1140–1154.

DE PLOEY, J. (1984): Hydraulics of runoff and loess loam deposition. Earth Surface Processes and Landforms, **9**, 533–539.

GOVERS, G. (1985): Selectivity and transport capacity of thin flows in relation to rill erosion. CATENA, **12**, 35–49.

HEEDE, B.H. (1974): Stages of development of gullies in Western United States of America. Zeitschrift für Geomorphologie, N.F. **18**, 260–271.

IMESON, A.C. & VERSTRATEN, J.M. (1982): An examination of the relationship between soil physical and chemical properties and the development of badlands in Morocco. In: Badland Geomorphology and Pipe Erosion. BRYAN, R.B. & YAIR, A. (Eds.). Geo Books, Norwich.

IMESON, A.C., KWAAD, F.J.P.M. & VERSTRATEN, T.M. (1982): The relatiosnhip of soil physical and chemical properties to the development of badlands in Morocco. In: Badland Geomorphology and Piping. BRIAN, R.B. & YAIR, A. (Eds.), 47–70, Geo Books, Norwich.

IRELAND, H.A., SHARPE, C.F.S. & EARGLE, D.H. (1939): Principles of gully erosion in the Piedmont of South Carolina. U.S. Dept. Agric. Tech. Bull., **633**, 143 pp.

MITCHELL, J.K. & BUBENZER, G.D. (1980): Soil loss estimation. In: Soil Erosion. KIRKBY, M.J. & MORGAN, R.P. (Eds.), Wiley, Chichester.

PIEST, R.F., BRADFORD, J.M. & SPOMER, R.G. (1975): Mechanics of erosion and sediment movement from gullies. Agric. Research Serv., USDA, ARS-S-40, 162–176.

SAVAT, J. (1976): Discharge velocities and total erosion of a calcareous loess; a comparison between pluvial and terminal runoff. Revue de Géomorphologie Dynamique, **24**, 113–122.

SAVAT, J. & DE PLOEY, J. (1982): Sheetwash and rill development by surface flow. In: Badland Geomorphology and Piping. BRYAN, R.B. & YAIR, A. (Eds.), 113–126. Geo Books, Norwich.

Address of author:
Jan de Ploey
Laboratory of Experimental Geomorphology
Redingenstraat 16 bis
B-3000 Leuven
Belgium

LONGITUDINAL VARIATIONS IN THE SIZE AND SORTING OF BED MATERIAL ALONG SIX ARID-REGION MOUNTAIN STREAMS

B.L. **Rhoads**, Urbana

Summary

The study uses regression analysis to investigate downstream patterns in mean grain size and sorting of bed material along six small ephemeral streams in a mountain watershed, southern Arizona, USA. Mean grain size declines exponentially along four of the six streams. The rates of decline are generally an order of magnitude greater than those reported previously for other streams throughout the world. These high rates reflect exceptionally rapid downstream changes in fluvial energy and sediment supply along these arid mountain streams. Downstream trends in sorting are more weakly developed than those for mean grain size; however, three of the streams exhibit improvements in the degree of sorting over distance. Mean grain size is significantly related to estimates of shear stress for all six streams, whereas the degree of sorting depends most strongly on the size of the bed material. Local variations in parent material also have significant influences on mean grain size and sorting within some of the streams.

ISSN 0722-0723
ISBN 3-923381-17-4
©1989 by CATENA VERLAG,
D-3302 Cremlingen-Destedt, W. Germany
3-923381-17-4/89/5011851/US$ 2.00 + 0.25

1 Introduction

Despite the fact that mountains occur in many dryland regions of the world, few geomorphological studies have investigated fluvial forms and processes within arid and semiarid mountain drainage systems. This situation probably reflects problems associated with obtaining accurate measurements of ephemeral fluvial events in remote, inaccessible environments under extremely harsh climatic conditions (SCHICK 1978). This study analyzes longitudinal patterns in bed material along six high-gradient ephemeral streams in the McDowell Mountains of southern Arizona. The research addresses the following questions:

1. Do the mean grain size and sorting of surficial bed materials change systematically along ephemeral mountain streams?, and

2. Do patterns in grain size and sorting reflect local variations in fluvial energy and sediment input?

Bed material is an important component of any fluvial system because it is an index of sediment transport and supply as well as a partial determinant of channel morphology, plan form, and gradient.

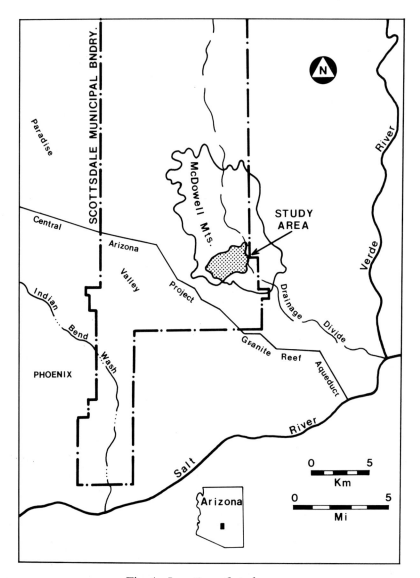

Fig. 1: *Location of study area.*

2 Field area

The study area is a small (6.8 km²) watershed located at the southern end of the McDowell Mountains in south-central Arizona (fig.1). The physical characteristics of these mountains are typical of those for ranges in the southwestern United States: the maximum elevation is 1240 m; the mean range relief is approximately 600 m; the adjacent basins have external drainage (i.e. they are not closed) and are large in area compared to the mountain mass; the range length and width are less

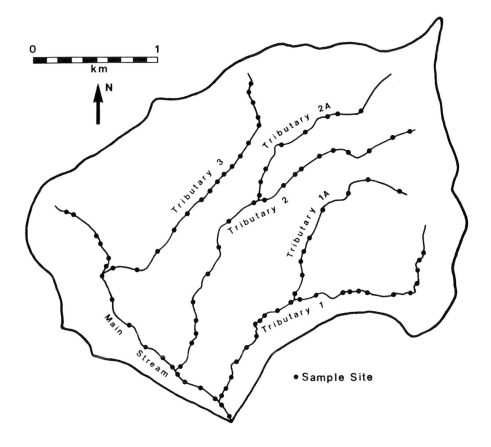

Fig. 2: *Distribution of sample sites along six streams in the study area.*

than 25 km and 5 km, respectively; the mountain front is dissected by small (1 to 10 km²) drainage basins; the range trends southeast-northwest; and the adjacent piedmont consists of erosionally-dissected pediments, alluvial fans, and bajadas (THORNBURY 1965, HUNT 1974). The climate of the mountains is arid subtropical with a mean annual temperature of 20°C and mean annual precipitation of 250–300 mm (CHRISTENSON et al. 1978). Vegetation consists of small trees, shrubs, and cacti (TURNER 1974). No altitudinal transitions in vegetation occur within the mountain range.

Six streams in the study basin were selected for analysis (fig.2). Each of these streams represents the trunk stream (i.e. the stream with the largest drainage area and lowest gradient at each confluence along its length) of the subnetwork above its mouth. All of the streams are ephemeral; analysis of hydrologic records for a nearby mountain basin indicates that geomorphically significant flows almost always occur as flash floods in response to intense convective storms (RHOADS 1986). In some years there may be no flows at all. Stream channel gradients range from 200 m/km near the

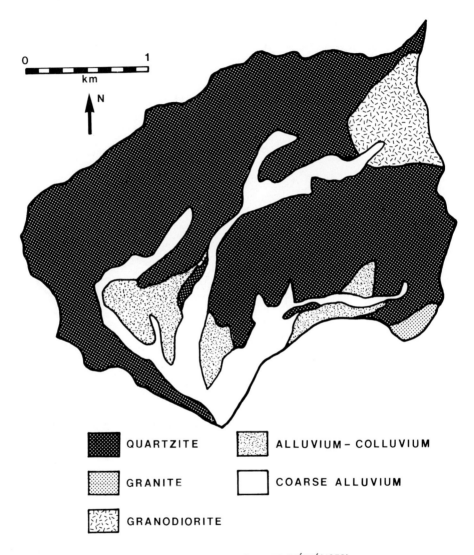

Fig. 3: *Geology of the study area.*

drainage divide to 20 m/km at the basin outlet. Headwater channels lie within narrow steep-sided (30–40°) bedrock valleys. Lower reaches incise coarse alluvium to form arroyos. The cause and timing of channel entrenchment in the McDowell Mountains are unknown (WELSCH 1977), but may be related to the development of basin-sited arroyos in the American Southwest between 1865 and 1900 (MELTON 1965). At many locations along each stream the arroyo banks are unstable as evidenced by detached blocks of calcified bank material.

Outcropping bedrock in the basin consists mainly of blocky or platy schistosic quartzite, which is the major sediment source for upper reaches of the streams (fig.3). Granitic intrusions occur locally at the eastern and southern ends of the watershed; however, only the massive, coarse-grained granodiorite is a significant sediment source, contributing large amounts of grus to the headwaters of Tributary 2 (figs. 2 & 3). Coarse alluvium is the primary sediment source for lower reaches of the channels. This material consists of quartzite boulders within a sand-silt matrix (CHRISTENSON et al. 1978).

3 Data

3.1 Field data

Bed material samples were collected at 94 sites along the six streams (fig.2). Five criteria governed the selection of these sites:

1. the sample reaches had streambeds consisting of alluvial material (i.e. bare bedrock channels were excluded);

2. they were relatively straight without significant expansions or contractions;

3. the sites were free of vegetation that might disrupt sediment transport;

4. reaches immediately upstream and downstream from tributary confluences were sampled unless they severely violated one of the other criteria; and

5. within the constraints of the other criteria the sites were spaced as evenly as possible along each stream.

One-hundred particles were sampled from the surface of the active streambed at each site using the grid-by-number method (WOLMAN 1954, KELLERHALS & BRAY 1971). The active streambed consisted of alluvial materials having a fresh appearance and lacking surficial weathering or varnish layers. The intermediate (b) axis of each particle was measured with either calipers (<50 mm) or a measuring tape (>50 mm). Mean grain size (graphic mean) and sorting (inclusive graphic standard deviation) of the samples were determined graphically from cumulative frequency curves (FOLK 1980) (tab.1). Surveys with an optical ranging device and Abney level provided data on channel gradients and cross-sectional profiles at the sample sites. Also, the composition and relative stability of the slopes flanking the channels (arroyo banks versus bedrock valley sides) were assessed qualitatively.

3.2 Estimation of shear stress

Consideration of the second research question posed at the beginning of this

		Mean	Standard Deviation	Min.	Max.	Skewness	Kurtosis
Main Stream (N = 18)	L	1.49	0.75	0.46	2.61	0.02	-1.57
	ln(L)	0.25	0.59	-0.79	0.96	-0.47	1.26
	$\overline{D}$	26.80	13.00	10.30	52.00	0.49	-0.39
	ln($\overline{D}$)	3.16	0.53	2.33	3.95	-0.31	-1.05
	σ	1.86	0.41	1.15	2.48	-0.13	-1.14
	τ_{25}	129.90	65.80	43.00	255.00	0.55	-0.51
	ln(τ_{25})	4.73	0.55	3.76	5.54	-0.38	-0.64
Tributary 1 (N = 23)	L	1.99	0.84	0.64	3.30	-0.12	-1.17
	ln(L)	0.58	0.50	-0.45	1.19	-0.75	-0.51
	$\overline{D}$	23.20	7.20	11.00	38.00	0.00	-0.71
	ln($\overline{D}$)	3.09	0.34	2.40	3.64	-0.55	-0.64
	σ	1.90	0.33	1.28	2.45	-0.51	-0.68
	τ_{25}	297.10	162.70	107.00	742.00	1.52	2.20
	ln(τ_{25})	5.57	0.49	4.67	6.60	0.36	-0.11
Tributary 1A (N = 9)	L	1.30	0.50	0.50	1.95	-0.49	-1.13
	ln(L)	0.18	0.47	-0.70	0.67	-0.99	-0.15
	$\overline{D}$	36.90	10.50	24.30	58.40	1.04	1.01
	ln($\overline{D}$)	3.57	0.27	3.19	4.06	0.51	-0.17
	σ	2.16	0.19	1.77	2.36	-1.32	1.03
	τ_{25}	405.10	210.10	121.00	804.00	0.90	0.40
	ln(τ_{25})	5.88	0.55	4.80	6.69	-0.49	1.19
Tributary 2 (N = 19)	L	2.11	1.00	0.61	3.75	0.20	-1.24
	ln(L)	0.62	0.54	-0.50	1.32	-0.53	-0.55
	$\overline{D}$	12.15	6.50	5.00	29.40	1.18	1.16
	ln($\overline{D}$)	2.37	0.51	1.61	3.38	0.28	-0.77
	σ	1.72	0.50	1.00	2.81	0.73	-0.23
	τ_{25}	320.40	193.10	102.00	735.00	0.83	-0.36
	ln(τ_{25})	5.60	0.60	4.62	6.60	0.15	-1.28
Tributary 2A (N = 8)	L	1.67	0.49	1.03	2.31	0.05	-1.95
	ln(L)	0.48	0.31	0.03	0.83	-0.19	-1.76
	$\overline{D}$	39.90	18.90	17.40	72.00	0.80	-0.37
	ln($\overline{D}$)	3.59	0.47	2.85	4.28	0.05	-0.54
	σ	2.25	0.33	1.69	2.70	-0.30	-0.49
	τ_{25}	574.00	239.70	241.00	879.00	-0.01	-1.52
	ln(τ_{25})	6.26	0.47	5.48	6.78	-0.58	-0.75
Tributary 3 (N = 17)	L	1.26	0.68	0.25	2.55	0.31	-0.63
	ln(L)	0.06	0.66	-1.37	0.93	-0.86	0.16
	$\overline{D}$	22.80	11.30	10.30	57.70	2.01	5.24
	ln($\overline{D}$)	3.04	0.42	2.33	4.05	0.65	0.87
	σ	1.91	0.46	1.36	2.94	0.76	-0.41
	τ_{25}	241.70	76.30	154.00	420.00	0.90	0.13
	ln(τ_{25})	5.44	0.30	5.04	6.04	0.76	-0.41

L = Distance Downstream (km); $\overline{D}$ = Mean Grain Size (mm); σ = Sorting Coefficient (phi units); τ_{25} = Shear stress of 25-year flood (N/m^2)

Tab. 1: *Descriptive statistics of variables for the six streams in the study area.*

paper requires a measure of fluvial energy. Several studies have demonstrated that mean boundary shear stress is related to flow competence (e.g. BAKER & RITTER 1974, CHURCH 1978, CARLING 1983). Longitudinal variations in competence should strongly influence downstream patterns in bed material characteristics.

Because no streamflow data are available for streams in the Mc-Dowell Mountains, values of shear stress were estimated using a two-stage procedure involving hydrologic and hydraulic modeling. The 25-year flood was selected as the design event for the modeling procedure because it generates shear stresses that are large enough to move the coarse bed material of the six streams. The first stage of the modeling procedure involved computing a discharge for the 25-year flood at each sample site using a distributed hydrologic model developed by LANE (1982). This model provides reasonable first-order approximations of discharges of various frequencies for small watersheds in southern Arizona (LANE 1982, 1985). It combines a modified version of the U.S. Soil Conservation Service curve numer method with a transmission loss function. Inputs to the model for each site include total, upland, and lateral drainage areas; length of the channel between the sample site and the site immediately upstream; the average width of the channelbed; estimates of runoff curve numbers for upland and lateral contributing areas; and the 25-year one-hour rainfall.

The discharges estimated by the distributed model represent hillslope runoff hydrographs modified by the transmission loss function. However, shear stress depends not only on discharge but also on the cross-sectional form and gradient of the stream channel. In the second stage of the modeling procedure, the computer algorithm DISCALC (RHOADS 1987) was used to estimate the flow characteristics (width, depth, velocity, hydraulic radius) of the 25-year flood for each site based on GOLUBSTOV'S (1969) resistance formula for steep mountain streams and the surveyed dimensions of the stream channel. The program calculated these values by iteratively adjusting the water surface elevation within a fixed cross section until the computed discharge converged on the estimate from the hydrologic model. At a few of the sites the channels were flanked by flat floodplains and iteration was terminated at the bankfull discharge because the maximum average shear stress for the channel-floodplain combination occurred at this depth of flow. The program computes shear stress (τ_{25}) as:

$$\tau_{25} = \gamma RS \qquad (1)$$

where R is the hydraulic radius of flow for the 25-year flood, S is the local gradient of the stream channel as surveyed in the field, and γ is the specific weight of the water sediment mixture (assumed to be 11,000 N/m^3 for sediment-laden floodwaters) (tab.1).

4 Downstream trends in grain size and sorting

The systematic downstream decline of particle size is a well-known phenomenon for alluvial streams (e.g. KNIGHTON 1984, 78–81). This trend is primarily the result of two processes associated with sediment transport:

1. hydraulic sorting, which involves the selective entrainment and differen-

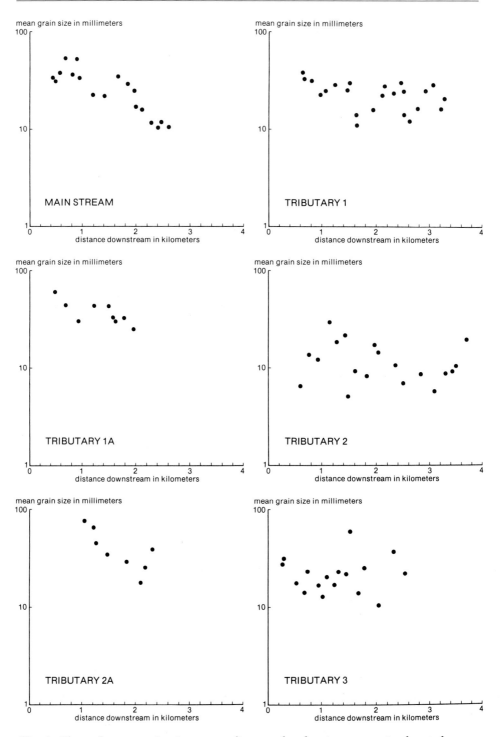

Fig. 4: *Plots of mean grain size versus distance for the six streams in the study area.*

Stream	ln(D₀)	a	SE	SEE	R²
Main Stream	4.07	-.61**	.09	.28	.74
Tributary 1	3.32	-.15⁺	.10	.32	.16
Tributary 1A	4.10	-.40*	.13	.19	.57
Tributary 2	2.65	-.13	.12	.51	.07
Tributary 2A	4.85	-.76*	.24	.31	.62
Tributary 3	3.00	.02	.16	.44	.00

** significant at the .01 level
* significant at the .05 level
⁺ significant at the .10 level
SE - standard error
SEE - standard error of estimate

Tab. 2: *Results of regression analyses — mean grain size versus distance.*

tial transportation of different particle sizes, and

2. abrasion, which refers to particle comminution by mechanical processes (grinding, rubbing, collision, **in situ** vibration).

Numerous field and laboratory studies have confirmed the appropriateness of an exponential model for describing the individual downstream effects of hydraulic sorting and abrasion (e.g. KRUMBEIN 1941, KUENEN 1956, CHURCH & KELLERHALS 1978, RANA et al. 1973, TROUTMAN 1980, KNIGHTON 1980):

$$D = D_0 e^{-aL} \qquad (2)$$

where D is the particle size (mm) at distance L (km), D_0 is the initial particle size at $L = 0$ (mm), and a is a coefficient summarizing the combined influences of hydraulic sorting and abrasion.

The basic data illustrating the relations among particle size and distance downstream from the drainage divide for the six streams in the McDowell Mountains are shown in fig.4. Because particle size is plotted on a logarithmic scale versus distance on an arithmetic scale, an exponential relationship will form a straight line on these diagrams. Scatterplots for the Main Stream and Tributaries 1, 1A, and 2A suggest a linear trend; however these data also exhibit considerable scatter. The other plots (Tributaries 2 and 3) display no apparent trends.

Statistical analyses confirm the conclusions based on visual inspections of the scatterplots (tab.2). The exponential model is statistically significant ($\alpha < 0.5$) for the Main Stream and Tributaries 1A and 2A, and marginally significant ($\alpha < .10$) for Tributary 1. The significant rate coefficients are order of magnitude greater than many previously published values (KNIGHTON 1987, Table 5.2). Similar rate coefficients are reported by MAYER et al. (1984) for Boulder Wash ($a = -.44$) and Stoval-Mohawk Wash ($a = -.34$), two ephemeral upland streams in southwestern Arizona. Their findings and the results in tab.2 indicate that exceptionally rapid changes in mean grain size occur along arid-region upland streams. These high rates probably reflect corresponding rates of change in fluvial energy and sediment supply.

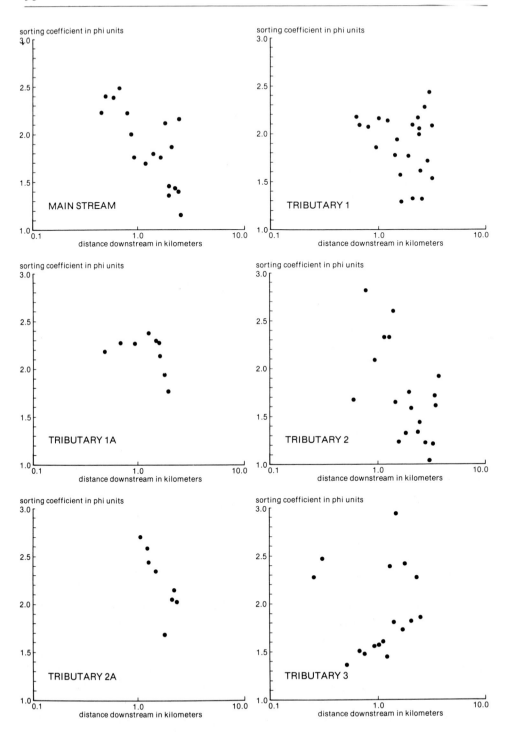

Fig. 5: *Plots of sorting coefficient versus distance for the six streams in the study area.*

Grain size is not significantly related to distance for Tributaries 2 and 3. Two factors appear to be responsible for the lack of decline in particle size along Tributary 2. The primary headward sediment source for this stream is fine-grained granodioritic grus. Field observations indicate that the size of this grus remains fairly constant as it is transported along this stream. The situation for Tributary 3 is less clear. The lack of diminution over distance may reflect suppressed delivery of coarse material to the headwaters of this stream due to the low relief in this portion of the basin. These interpretations for Tributaries 2 and 3 are supported by the relatively small values for D_0 in their corresponding regression equations (tab.2), which confirm that the initial size of the parent material is not as large for these streams as it is for the others. In fact, the correlation between D_0 and the rate coefficients (a's) for the six equations is -.93, supporting the contentions of KRUMBEIN (1941) and KUENEN (1956) among others, that the rate of decline in particle size depends on the initial size of the sediment supplied to the headwaters of a stream.

Previous research provides little theoretical insight into the relationship between the degree of sorting and distance. Physically-based and stochastic models of longitudinal sediment transport indicate that sorting should improve downstream (e.g. RANA et al. 1973, TROUTMAN 1980, MCLAREN & BOWLES 1985). However, the field evidence is inconclusive. MCPHERSON (1971) and SMITH (1970, 1974) reported downstream improvement in the degree of sorting, but other investigators found no obvious trend in the magnitude of the sorting coefficient over distance (KRUMBEIN 1940, 1942, PLUMLEY 1948, POTTER 1955, MILLER 1958, INDERBITZEN 1959). KNIGHTON (1980) observed cyclic patterns in sorting along several streams in Britain, which he attributed to disruptions by tributary inputs of a general downstream trend for improved sorting.

In order to determine the existence and form of a possible relationship between sorting and distance for the six streams in the McDowell Mountains, all possible combinations of log-transformed and untransformed values of these two variables were plotted on bivariate diagrams. These plots revealed that the logarithmic model:

$$\sigma = \sigma_0 + b ln(L) + \varepsilon \qquad (3)$$

where σ is the sorting coefficient (in phi units), σ_0 is the initial sorting at $ln(L) = 0$, and ε is an error term, is appropriate for all of the streams except Tributary 1A (fig.5). The data for this stream were also not linear on any of the other scatterplots. In general, longitudinal trends in sorting are more poorly developed than those for mean grain size, but downstream improvements in the degree of sorting are evident for the Main Stream, and Tributaries 2 and 2A (fig.5). No periodic variations in the magnitude of the sorting coefficient, such as those described by KNIGHTON (1980), are apparent on any of the scatterplots.

As with mean grain size, statistical analyses confirm the visual observations (tab.3). Tributary 2A, which has the highest rate of downstream decline in particle size, also has the largest rate coefficient (b) for degree of sorting. Rates of improvement in sorting for the Main Stream and Tributary 2 are almost identical despite the fact that, as mentioned above, mean particle size declines along the former but remains fairly constant

Stream	σ_0	b	SE	SEE	R^2
Main Stream	1.99	-.52**	.11	.27	.57
Tributary 1	1.97	-.13	.14	.33	.04
Tributary 1A		nonlinear			
Tributary 2	2.06	-.54**	.18	.42	.34
Tributary 2A	2.67	-.89*	.25	.21	.67
Tributary 3	1.90	.04	.18	.48	.00

** significant at the .01 level
* significant at the .05 level

Tab. 3: *Results of regression analyses — sorting versus distance.*

along the latter. As in the equations for mean grain size, the degree of association between the constant term (σ_0) and the rate coefficient (b) is strong for the sorting-distance models ($R^2 = .70$). Apparently, the downstream rate of improvement in sorting is inversely related to the degree of sorting of the bed material in headwater reaches.

5 Factors controlling longitudinal variations in mean grain size and sorting

Longitudinal variations in the mean grain size and sorting of bed material should reflect spatial variations in characteristics of material contributed to the channel and modification of this parent material by sediment transport. In order to test this hypothesis for mean grain size, the following general model was formulated:

$$ln(\overline{D}) = ln(D_0) + b_1\tau_{25} + b_2X_2 + ln(\varepsilon) \quad (4)$$

where τ_{25} is the estimated mean boundary shear stress of the 25-year flood (N/m²), and X_2 is a dummy variable representing the type of bank material at each site (0 for arroyos, 1 for channels flanked by bedrock valleysides). The model was expanded for the Main Stream to include an additional dummy variable, x_3, which measures the influence of fine-grained grus contributed by Tributary 2 on bed material characteristics of the Main Stream below its confluence with this tributary (0 for sites upstream from the confluence, 1 for sites downstream from the confluence). Thus the Main Stream model has the form:

$$ln(D) = ln(D_0) + b_1\tau_{25} + b_2X_2 + b_3X_3 + ln(\varepsilon) \quad (5)$$

Backward elimination stepwise regression was employed to estimate the model for each stream. Independent variables were deleted from the fully-specified model one at a time until the remaining variables produced F ratios significant at the .10 level. The F ratio measures the incremental contribution of each variable to the explained sum of squares, assuming all of the remaining variables are already included in the regression.

The average explained variation (R^2) for the resulting models is rather low (.48), but it represents an improvement over the grain size-distance models (average $R^2 = .36$) (tab.4). The estimated shear stress of the 25-year flood remained in all six equations as at least a marginally significant variable, including those for Tributaries 2 and 3 which

Stream	ln(D_0)	b_1	SE	b_3	SE	b_3	SE	SEE	R^2
Main Stream	3.87	.0021* (.26)	.0010	–	–	-.83** (-.77)	.13	.24	.82
Tributary 1	3.41	.0012**	.0004	–	–	NA	NA	.28	.35
Tributary 1A	3.18	.0011** (.87)	.0003	-.53* (-.66)	.21	NA	NA	.17	.69
Tributary 2	2.03	.0011+	.0006	–	–	NA	NA	.48	.16
Tributary 2A	2.82	.0013+	.0006	–	–	NA	NA	.38	.46
Tributary 3	2.36	.0034* (.62)	.0011	-.33+ (-.41)	.18	NA	NA	.35	.40

** significant at the .01 level
* significant at the .05 level
+ significant at the .10 level
- not significant at the .10 level
NA - Not applicable
standardized coefficients shown in parentheses

Tab. 4: *Results of stepwise regression analyses — mean grain size versus shear stress, type of bank material, and lithology.*

do not exhibit significant relationships between mean grain size and distance (tab.2). The values for b_1 indicate that a positive relationship exists between shear stress and grain size, i.e. mean grain size is largest where flow competence is greatest. These results confirm that longitudinal patterns in grain size are strongly influenced by downstream variations in fluvial energy.

Variations in sediment supply appear to have a significant influence on mean grain size within at least three of the streams. Negative values for b_2 in the regression equations for Tributary 1A and Tributary 3 reveal that reaches of these streams flanked by bedrock valleysides tend to have smaller mean grain sizes than those within arroyos. Field observations indicated that weathered quartzite at bedrock outcrops consists of relatively small platy particles, whereas the coarse alluvium comprising the arroyo banks contains large quartzite boulders that are probably products of a past weathering regime. The contribution of these boulders to the channels at arroyo cutbanks locally increases the mean grain size. Results also show that the addition of fine-grained granodioritic grus from Tributary 2 to lower reaches of the Main Stream leads to a reduction in mean grain size. In fact, the standardized values of b_1 and b_2 for the Main Stream equation indicate that mean grain size is more strongly affected by this contributed material than by variations in shear stress.

Factors influencing the degree of sorting along each stream were investigated using the model:

$$\sigma = \sigma_0 + b_1 ln(\tau_{25}) +$$

Stream	σ_0	b_1	SE	b_2	SE	b_3	SE	b_4	SE	SEE	R^2
Main Stream	.22	–	–	–	–	–	–	.51**	.14	.31	.45
Tributary 1	.04	–	–	–	–	NA	NA	.60**	.17	.26	.38
Tributary 1A	.75	–	–	–	–	NA	NA	.40+	.21	.17	.31
Tributary 2	-1.75	.43** (.52)	.12	-.30+ (-.25)	.17	NA	NA	.47** (.47)	.15	.30	.71
Tributary 2A	-.77	.49** (.70)	.11	-.30* (-.27)	.11	NA	NA	–	–	.12	.90
Tributary 3	-.37	–	–	-.24+ (-.27)	.14	NA	NA	.79** (.72)	.17	.28	.68

** significant at the .01 level
* significant at the .05 level
+ significant at the .10 level
- not significant at the .10 level
NA - not applicable
standardized coefficients shown in parentheses

Tab. 5: *Results of stepwise regression analyses — sorting versus shear stress, type of bank material, lithology, and mean grain size.*

$$b_2 ln(X_2) + b_4 ln(\overline{D}) + \varepsilon \quad (6)$$

Mean grain size $(\overline{D})$ is included as an explanatory variable representing the relative inertia of the bed material. As explained above, the dummy variable X_3 was included in the model for the Main Stream only:

$$\sigma = \sigma_0 + b_1 ln(\tau_{25}) + b_2 ln(X_2) + b_3 ln(X_3) + b_4 ln(\overline{D}) + \varepsilon \quad (7)$$

For computational convenience, values of X_2 and X_3 were set at 1.0 and 2.72 so that logarithmically-transformed values became 0 and 1, respectively. The sorting models were estimated using backward-elimination stepwise regression.

The estimated sorting models account for 54% of the total variance for the six streams compared to an average R^2 of 26% for the sorting distance models (tab. 3 & 5). Mean grain size is the dominant explanatory variable, remaining in five of the six equations. At least two physical processes may be responsible for this strong association between particle size and sorting. First, the size distribution for relatively coarse bed material will not be modified as easily by streamflow as will the distribution for relatively fine material. In other words, hydraulic sorting will operate less efficiently on the former than on the later. Second, flow separation caused by coarse particles (cobbles, boulders) tends to increase the variance of the particle size distribution by restricting the entrainment and promoting the deposition of fine grains (LARONNE & CARSON 1976, BRAYSHAW 1985). It is also possible that this relationship reflects a statistical dependence between the magnitudes of sorting (standard deviation) and mean grain size. Such a relationship is common for arithmetic measures of mean and standard deviation. However, FOLK (1980) shows that the relationship between sorting (spread of the distribution in phi units) and mean grain size (geometric mean) is not inherently linear for fluvial sediments.

Mean grain size did not remain in the regression equation for Tributary 2A. Instead, shear stress is the dominant explanatory variable for this stream. This result suggests that along this steep tributary shear stress (force/area) may have a greater influence on sorting than does mean grain size (inertia, infiltration of fines). On the other hand, the exclusion of mean grain size may be due to fortuitously stronger covariance between stress and sorting than between mean grain size and sorting in this small sample ($N = 8$).

All three independent variables remained in the model for Tributary 2. The standardized coefficients indicate that shear stress is slightly more influential than mean grain size, although their effects are roughly equal. The significance of shear stress probably reflects the high mobility of the relatively fine-grained granodioritic grus within Tributary 2. The effect of bank material is only marginally significant and relatively weak compared to the influences of particle size and shear stress. The negative values for the bank material coefficient (b_2) in this model and in the models for Tributaries 2A and 3 suggest that weathered bedrock parent material is better sorted than the coarse alluvium contributed to the channels at arroyo cutbanks.

6 Effects of tributary inputs on mean grain size and sorting

The residuals of the regression analyses described above represent variation in mean grain size and sorting that is unaccounted for by the estimated models. These residuals have two components: stochastic error, which encompasses effects on the dependent variable of independent variables not included in the analysis, and measurement error. If it is assumed that the latter type of error is minor relative to the former, the residuals should primarily reflect unspecified influences on mean grain size and sorting. One factor that is frequently mentioned as a potentially significant source of residual variation for grain size and sorting is the input of matrial to a stream by tributaries (e.g. RICHARDS 1982, 232, KNIGHTON 1984, 80, PETTS & FOSTER 1985, 179). Results of field research (CHURCH & KELLERHALS 1978, KNIGHTON 1980) and stochastic simulations (TROUTMAN 1980) indicate that both mean grain size and the sorting coefficient increase at junctions as a result of tributary sediment contributions. These findings imply that regression residuals for sites at channel junctions should be positive.

The sample sites for this study represent two groups:

1. tributary sites, or those situated immediately downstream (<30 m) from channel junctions ($N = 24$), and

2. intertributary sites, or those located between junctions ($N = 70$).

The signs of the regression residuals for the two groups were compared using the Chi square test. Results reveal

a) Residuals from mean grain size regression analyses (tab.5)
Null hypothesis: No significant differences between types of sites with respect to the signs of the residuals

Sign	Tributary	Intertributary	Total
Positive	15 (12.3)	33 (35.7)	48
Negative	9 (11.7)	37 (34.3)	46
Total	24	70	94

df = 1 $\chi^2 = 1.62$ χ^2 at α .10 = 2.71

Decision: Cannot reject null hypothesis

b) Residuals from sorting regression analyses (tab.6)
Null hypothesis: No significant differences between types of sites with respect to the signs of the residuals

Sign	Tributary	Intertributary	Total
Positive	10 (11.7)	36 (34.3)	46
Negative	14 (12.3)	34 (35.7)	48
Total	24	70	94

df = 1 $\chi^2 = .64$ χ^2 at $\alpha = .10 = 2.71$

Decision: Cannot reject null hypothesis

Tab. 6: *Chi square test — signs of regression residuals for tributary versus intertributary sites.*

that the mean grain size residuals for tributary sites show a slight tendency to be positive, but tributary and intertributary residuals do not differ significantly for mean grain size or for sorting (tab.6). This finding suggests that either tributary inputs do not strongly affect bed material characteristics in this drainage network or that the streams undergo local adjustments (primarily via shear stress) to accomodate the introduced load. Alternatively, systematic influences by tributaries may be obscured by measurement error or by local contributions of sediment between the junctions.

7 Conclusion

This paper has attempted to explain longitudinal patterns in mean grain size and sorting of bed material along six aridregion mountain streams. Distance is only marginally useful as an explanatory variable since several of the streams exhibit little or no systematic longitudinal variation in mean grain size and/or sorting. This finding corresponds with results of other studies of downstream patterns in bed material characteristics along upland streams (HACK 1957, BRUSH 1961, MILLER 1958). However, the rate coefficients that are statistically significant are relatively high compared to those reported previously, indicating that exceptionally rapid changes in mean grain size and sorting occur in these high-gradient streams. The initial size and sorting of material supplied to the headwaters of the channels appears to determine the magnitude of these rates, a phenomenon that has been endorsed by BLUCK (1987).

The level of explanation improves for models that directly incorporate variables representing fluvial energy and sediment supply, but the amount of unexplained variation is still relatively large ($\approx$ 50%). The strong relationship between mean grain size and shear stress implies that longitudinal variations in particle size primarily reflect spatial changes in flow competence. Local increases in shear stress result in the movement of larger particles as well as more effective downstream flushing (selective transport) of fine material. Both of these processes tend to increase mean grain size. Field observations indicate that selective winnowing of the finer grain-size fractions may be primarily responsible for longitudinal variations in mean grain size, especially in headwater reaches where the interlocking structure of boulder steps retards the movement of coarse particles. Sorting on the other hand is strongly associated with mean grain size. This relationship may reflect two processes:

1. increasing ineffectiveness of flows to sort bed material as its size increases or

2. infiltration of fines due to disruption of flow by coarse bed material.

Undoubtedly, some of the residual variation in the statistical models is due to measurement or sampling error, particularly for mean grain size, sorting, and the estimated shear stress. However, a high degree of unexplained variation in particle size-distance relationships appears to be a general characteristic of mountain streams and has been attributed to abundant local sediment sources (MCPHRESON 1971). Significant systematic infuences by one type of local sediment source—tributary inputs—were not detected in this study. Another factor that may contribute to unexplained variation is small-scale temporal and spatial variability of bedload transport, deposition, and storage, perhaps in the form of waves (e.g. LEKACH & SCHICK 1983, MEADE 1985). In ephemeral mountain streams, bedload transport by spatially varied unsteady flows of flash floods probably enhances the wavelike movement and longitudinal variability of bed material. Also, bedforms may exert a strong influence on downstream changes in bed material in high-gradient, coarse-bedded streams by controlling selective clast retention and rejection (BLUCK 1987). More microscale (within-reach) analyses of bed material variegation are required to identify sources of residual variation and explain unsystematic downstream patterns in mean grain size and sorting. Ultimately, it is at this scale of analysis that studies of bed material characteristics can make the greatest contributions to theories concerning sediment transport and stream channel morphology.

References

BAKER, V.R. & RITTER, D.F. (1975): Competence of rivers to transport coarse bedload material. Geological Society of America Bulletin, **86**, 975–978.

BLUCK B.J. (1987): Bed forms and clast size changes in gravel-bed rivers. In: River Channels: Environment and Process. (Ed. K.S. RICHARDS), Basil Blackwell, New York, 159–178.

BRAYSHAW, A.C. (1985): Bed microtopography and entrainment thresholds in gravel-bed rivers. Geological Society of America Bulletin, **96**, 218–223.

BRUSH, L.M. (1961): Drainage basins, channels, and flow characteristics of selected streams in central Pennsylvania. U.S. Geological Survey Professional Paper 282-F.

CARLING, P.A. (1983): Threshold of coarse sediment transport in broad and narrow natural streams. Earth Surface Processes and Landforms, **8**, 1–18.

CHRISTENSON, G.E., WELSCH, D.G. & PÈWÈ, T.L. (1978): Environmental geology of the McDowell Mountains area, Maricopa County, Arizona. Folio Series, Map GI-1-A (Geology), Bureau of Mineral Technology and University of Arizona, Tucson, scale 1:24000.

CHURCH, M. (1978): Paleohydrological reconstruction from a Holocene valley fill. In: Fluvial Sedimentology. (Ed. A.D. MIALL), Canadian Society of Petroleum Geologists, 743–772.

CHURCH, M. & KELLERHALS, R. (1978): On the statistics of grain size variation along a gravel river. Canadian Journal of Earth Sciences, **15**, 1151–1160.

FOLK, R.L. (1980): Petrology of Sedimentary Rocks. Hemphill, Austin, Texas.

GOLUBSTOV, V.V. (1969): Hydraulic resistance and formula for computing the average velocity of mountain rivers. Soviet Hydrology: Selected Papers, **5**, 500–511.

HACK, J.T. (1957): Studies of longitudinal stream profiles in Virginia and Maryland. U.S. Geological Survey Professional Paper 294-B.

HUNT, C.B. (1974): Natural Regions of the United States and Canada. Freeman, San Francisco.

INDERBITZEN, A.L. (1959): Gravels of Alameda Creek, California. Journal of Sedimentary Petrology, **29**, 212–220.

KELLERHALS, R. & BRAY, D.J. (1971): Sampling procedures for coarse fluvial sediments. Journal of The Hydraulics Division, American Society of Civil Engineers, **97**, 1165–1180.

KNIGHTON, A.D. (1980): Longitudinal changes in size and sorting of stream-bed material in four English rivers. Geological Society of America Bulletin, **91**, 55–62.

KNIGHTON, A.D. (1984): Fluvial Forms and Processes. Edward Arnold, London.

KNIGHTON, A.D. (1987): River channel adjustment — the downstream dimension. In: River Channels: Environment and Process. (Ed. K.S. RICHARDS), Basil Blackwell, New York, 95–128.

KRUMBEIN, W.C. (1940): Flood gravel of San Gabriel Canyon. Geological Society of America Bulletin, **51**, 639–676.

KRUMBEIN, W.C. (1941): The effects of abrasion on the size and shape of rock fragments. Journal of Geology, **49**, 482–520.

KRUMBEIN, W.C. (1942): Flood deposits of Arroyo Seco, Los Angeles County, California. Geological Society of America Bulletin, **53**, 1355–1402.

KUENEN, P.H. (1956): Experimental abrasion of pebbles. 2. Rolling by current. Journal of Geology, **64**, 336–368.

LANE, L.J. (1982): Distributed model for small semi-arid watersheds. Journal of the Hydraulics Division, American Society of Civil Engineers, **108**, 1114–1131.

LANE, L.J. (1985): Estimating transmission losses. In: Development and Management Aspects of Irrigation and Drainage Systems, Proceedings of the Specialty Conference, Irrigation and Drainage Division, American Society of Civil Engineers, San Antonio, TX, 106–113.

LARONNE, J.B & CARSON, M.A. (1976): Interrelationships between bed morphology and bed-material transport for a small, gravel-bed channel. Sedimentology, **23**, 67–85.

LEKACH, J. & SCHICK, A.P. (1983): Evidence for transport of bedload in waves: analysis of fluvial samples in a small upland stream channel. CATENA, **10**, 276–279.

MAYER, L., GERSON, R. & BULL, W.B. (1984): Alluvial gravel production and deposition: a useful indicator of Quaternary climatic change in deserts. CATENA SUPPLEMENT, **5**, 137–151.

MCLAREN, P. & BOWLES, D. (1985): The effects of sediment transport on grain-size distributions. Journal of Sedimentary Petrology, 55, 457–470.

MCPHERSON, H.C. (1971): Downstream changes in sediment character in a high energy mountain stream channel. Arctic and Alpine Research, 3, 65–79.

MEADE, R.H. (1985): Wavelike movement of bedload sediment, East Fork River, Wyoming. Environmental Geology and Water Science, 4, 215–225.

MELTON, M.A. (1965): The geomorphic and paleoclimatic significance of alluvial deposits in southern Arizona. Journal of Geology, 73, 1–38.

MILLER, J.P. (1958): High mountain streams; effects of geology on channel characteristics and bed material. New Mexico State Bureau of Mines and Mineral Resources Memoir, 4, 1–53.

PETTS, G. & FOSTER, I. (1985): Rivers and Landscape. Edward Arnold, London.

PLUMLEY, W.J. (1948): Black Hills terrace gravels: a study in sediment transport. Journal of Geology, 56, 526–577.

POTTER, P.E. (1955): The petrology and origin of the Lafayette gravel. Part 1: mineralogy and petrology. Journal of Geology, 63, 1–38.

RANA, S.A., SIMONS, D.B. & MAHMOOD, K. (1973): Analysis of sediment sorting in alluvial channels. Journal of the Hydraulics Division, American Society of Civil Engineers, 99, 1967–1980.

RHOADS, B.L. (1986): Process and response in desert mountain fluvial systems. Arizona State University, Ph.D. thesis.

RHOADS, B.L. (1987): DISCALC: A computer algorithm for computing the flow characteristics of flood discharges in stream channel cross sections. Computers and Geosciences, 13, 495–511.

RICHARDS, K.S. (1982): Rivers: Form and Process in Alluvial Channels. Methuen, New York.

SCHICK, A.P. (1978): Field experiments in arid fluvial environments: considerations for research design. Zeitschrift für Geomorphologie Supplementband, 29, 22–28.

SMITH, N.D. (1970): The braided stream depositional environment: comparison of the Platte River with some clastic rocks, north central Appalachians. Geological Society of America Bulletin, 81, 2993–3014.

SMITH, N.D. (1974): Sedimentology and bar formation in the Upper Kicking Horse River, a braided outwash stream. Journal of Geology, 82, 205–223.

THORNBURY, W.D. (1965): Regional Geomorphology of the United States. Wiley, New York.

TROUTMAN, B.M. (1980): A stochastic model for particle sorting and related phenomena. Water Resources Research, 16, 65–76.

TURNER, R.M. (1974): Map showing vegetation in the Phoenix area, Arizona. U.S. Geological Survey Miscellaneous Investigation Map I-845-I. 1:250000.

WELSCH, D.G. (1977): Environmental geology of the McDowell Mountains area, Maricopa County, Arizona: Part II. Arizona State University, M.S. thesis.

WOLMAN, M.G. (1954): A method of sampling coarse bed material. Transactions of the American Geophysical Union, 35, 951–956.

Address of author:
B.L. Rhoads
Department of Geography
University of Illinois
Urbana, IL 61801
USA

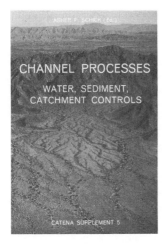

Asher P. Schick (Ed.):

CHANNEL PROCESSES
WATER, SEDIMENT, CATCHMENT CONTROLS

CATENA SUPPLEMENT 5, 1984

Price DM 110,—

ISSN 0722–0723 / ISBN 3-923381-04-2

PREFACE

Two decades ago, the publication of 'Fluvial Processes in Geomorphology' brought to maturity a new field in the earth sciences. This field – deeply rooted in geography and geology and incorporating many aspects of hydrology, climatology, and pedology – is well served by the forum provided by CATENA. Much progress has been accomplished in fluvial geomorphology during those twenty years, but the highly complex and delicate relationships between channel processes and catchment controls still raise intriguing problems. Concepts dealing with thresholds and systems, and modern tools such as remote sensing and sophisticated tracing, have not decisively resolved the simple but elusive dual problem: how does the catchment shape the stream channel and valley to its form, and why? And: how does the channel transmit its influence upstream in order to make the catchment what it is?

Partial solutions, in a regional or thematic sense, are common and important. In addition to contributing a building block to the study of fluvial geomorphology, they also produce a number of new questions. The consequent proliferation of research topics characterises this collection of papers. The basic tool of geomorphological interpretation – the magnitude, frequency, and mechanism of sediment and water conveyance – is a prime focus of interest. Increasingly important in this context in recent years is the role of human interference natural fluviomorphic process systems. Effects of drainage ditching, transport of pesticides absorbed in fluvial sediment, and the flushing of nutrients are some of the Manconditioned aspects mentioned in this volume. Other contributions deal with the intricate balance, especially in extreme climatic zones, between physical process generalities and macroregional morphoclimatic influences.

The contributions of PICKUP and of PICKUP & WARNER represent two of the very few detailed quantitative geomorphological analyses of very humid tropical catchments. The 8 to 10 m mean annual rainfall in the equatorial mountain areas studied combines with effective landsliding to produce extremely high denudation rates. However, many aspects of channel behaviour are similar to those of temperate rivers. Particularly interesting are the relationships derived between channel characteristics, perimeter sediment and bedload transport.

Several small ephemeral and intermittent streams in Ohio studied by THARP, although variable in catchment area and in peak discharge, have a similar competence; while sorting increases downstream, the coarsest sizes tend to remain constant. Sorting of fluvial sediment, though on a much longer time scale and in an arid climate, also plays an important role in the contribution MAYER, GERSON & BULL. They find that modern channel sediment size exhibits the most rapid downstream decrease in mean particle size, while Pleistocene deposits show the least rapid decrease and are consistently finer than young deposits. The difference is attributed to climatic change and a predictive mo thereto is presented.

HASSAN, SCHICK & LARONNE describe a new method for the magne tracing of large bedload particles capable of detecting tagged particles redeposi by floods up to several decimetres below the channel bed surface. Their meth may considerably enhance the value of numerous experiments with pain pebbles, previously reported or currently in progress. Suspended sediment is subject of the paper by CARLING. He experiments with sampling gravel-bedd flashy streams by two methods, and concludes that pump-sampling and 'buck sampling show significant differences only for very high discharges. Suspen sediment concentration is also dealt with by GURNELL & FENN, but in a prog cial environment – a climatic zone about which our knowledge is largely defici They find some correspondence between 'englacial' and 'subglacial' flow comp nents and the total suspended sediment concentration.

The effects of human interference by ditching in a forest catchment on s ment concentration and sediment yield is discussed by BURT, DONOHOE VANN. A local reservoir afforded an opportunity to monitor in detail the in ence of these drainage operations on the sediment concentration which incr sed dramatically and, after several months, gradually recovered due to revege tion. TERNAN & MURGATROYD analyse sediment concentrations and spe fic conductance in a humid, forest and marsh environment. Permanent vegetat dams are found to influence sediment concentration directly through filtrat and indirectly through changes in water depth and velocity. Changes in spe fic conductance are influenced by marsh inputs as well as by the addition of a of coniferous forest. The relationship between quality of water and fluvial s ment characteristics is dealt with by HERRMANN, THOMAS & HÜBNER, analyse the regional pattern of estuarine transport processes. They conclude high pesticide concentrations are correlated with high concentrations of susp ded sediment. Hydrodynamic rather than physicochemical factors influence regional distribution in the estuary, and the effect of brooklets draining intensiv cultivated land is quite evident.

Asher P. Schick

CONTENTS

G. PICKUP
*GEOMORPHOLOGY OF TROPICAL RIVERS
I. LANDFORMS, HYDROLOGY AND SEDIMENTATION IN THE FLY AND LOWER PURARI, PAPUA NEW GUINEA*

G. PICKUP & R. F. WARNER
*GEOMORPHOLOGY OF TROPICAL RIVERS
II. CHANNEL ADJUSTMENT TO SEDIMENT LOAD AND DISCHARGE IN THE FLY AND LOWER PURARI, PAPUA NEW GUINEA*

P. A. CARLING
COMPARISON OF SUSPENDED SEDIMENT RATING CURVES OBTAINED USING TWO SAMPLING METHODS

J. L. TERNAN & A. L. MURGATROYD
THE ROLE OF VEGETATION IN BASEFLOW SUSPENDED SEDIMENT AND SPECIFIC CONDUCTANCE IN GRANITE CATCHMENTS, S. W. ENGLAND

T. P. BURT, M. A. DONOHOE & A. R. VANN
CHANGES IN THE SEDIMENT YIELD OF A SMALL UPLAND CATCHMENT FOLLOWING A PRE-AFFORESTATION DITCHING

R. HERRMANN, W. THOMAS & D. HÜBNER
ESTUARINE TRANSPORT PROCESSES OF POLYCHLORINATED BIPHENYLS AND ORGANOCHLORINE PESTICIDES – EXE ESTUARY, DEVON

W. SEILER
MORPHODYNAMISCHE PROZESSE IN ZWEI KLEINEN EINZU GEBIETEN IM OBERLAUF DER ERGOLZ – AUSGELÖST DUF DEN STARKREGEN VOM 29. JULI 1980

A. M. GURNELL & C. R. FENN
FLOW SEPARATION, SEDIMENT SOURCE AREAS AND SUSP DED SEDIMENT TRANSPORT IN A PRO-GLACIAL STREAM

T. M. THARP
SEDIMENT CHARACTERISTICS AND STREAM COMPETENCE EPHEMERAL AND INTERMITTENT STREAMS, FAIRBORN, OHI

L. MAYER, R. GERSON & W. B. BULL
ALLUVIAL GRAVEL PRODUCTION AND DEPOSITION – A USE. INDICATOR OF QUATERNARY CLIMATIC CHANGES IN DESE (A CASE STUDY IN SOUTHWESTERN ARIZONA)

M. HASSAN, A. P. SCHICK & J. B. LARONNE
THE RECOVERY OF FLOOD-DISPERSED COARSE SEDIME PARTICLES – A THREE-DIMENSIONAL MAGNETIC TRAC METHOD

TALUS AND PEDIMENT FLATIRONS EROSIONAL AND DEPOSITIONAL FEATURES ON DRYLAND CUESTA SCARPS

K.-H. Schmidt, Berlin

Summary

Dryland cuesta scarps have been investigated on the Colorado Plateau in the southwestern United States and in the Pre-Saharan depression in southern Morocco. Talus and pediment flatirons on scarp slopes are the result of transport, depositional and erosional processes. The flatirons have a triangular to trapezoidal shape with their top directed towards the cuesta scarp, they are detached from the active scarp slope. Their backslope is protected by a thin caprock derived talus cover.

For the development of flatirons, two principal models have been offered; the non-cyclic model, which states that the flatirons have been formed within a climatically undifferentiated phase, and the cyclic model, which states that they are the product of climatically distinct phases. Though both regions have experienced climatic change from pluvial to interpluvial periods, cyclic forms are only predominant in southern Morocco. There are controlling factors of lithological and structural nature which favour the climatically dependent development of talus and pediment flatirons. These are thick lower slope rocks of homogeneous resistance (1), resistant caprocks preferably with carbonate cement (2), thickness ratios between 1 and 3 (3), the scarps should not be complex (4), and no subsequent valleys or deeply incised canyons should be in the scarp foreland (5). Conditions in southern Morocco are very favourable for cyclic flatiron formation. If flatiron sequences of different ages are found in the scarp foreland, they can be used to estimate rates of scarp retreat.

Zusammenfassung

Trockengebietsschichtstufen wurden auf dem Colorado Plateau im Südwesten der Vereinigten Staaten und in der Präsaharischen Senke in Südmarokko untersucht. Als Produkt von Transport-, Akkumulations- und Erosionsprozessen haben sich auf den Stufenhängen Schuttrampen und Fußflächenrampen gebildet. Die Rampen haben eine dreiecks- bis trapezförmige Gestalt, ihre Spitze ist gegen den Stufenhang gerichtet. Sie sind vom Stufenhang abgetrennt und tragen eine dünne vom Stufenbildner stammende Schuttschicht.

Für die Entstehung von Rampen sind zwei unterschiedliche Modelle entwickelt worden, das nicht-zyklische Modell, das annimmt, daß die Rampen

ISSN 0722-0723
ISBN 3-923381-17-4
©1989 by CATENA VERLAG,
D-3302 Cremlingen-Destedt, W. Germany
3-923381-17-4/89/5011851/US$ 2.00 + 0.25

innerhalb einer undifferenzierten klimatischen Phase geschaffen werden, und das zyklische Modell, das in den Rampen das Produkt eines Klimawechsels von feuchteren zu trockeneren Bedingungen sieht. Obwohl beide Gebiete Klimawechsel während des jüngeren Quartärs erfahren haben, sind die zyklischen Formen nur in Südmarokko weit verbreitet. Bestimmte lithologische und strukturelle Steuerungsfaktoren fördern die klimaabhängige Entwicklung von Schutt- und Fußflächenrampen. Dieses sind mächtige Sockelgesteine homogener Resistenz (1), resistente Stufenbildner möglichst mit Karbonatzement (2), eine Mächtigkeitsrelation zwischen 1 und 3 (3), die Stufenhänge sollten nicht komplex sein (4), und die Stufe darf nicht von subsequenten Gewässern oder tief eingeschnittenen Canyons beeinflußt sein (5). In Südmarokko sind die Bedingungen für die zyklische Rampenentwicklung sehr günstig. Wenn im Stufenvorland Abfolgen von verschieden alten Rampengenerationen vorhanden sind, können diese zur Bestimmung von Stufenrückwanderungsraten herangezogen werden.

1 Introduction

Dryland cuesta scarps have been investigated by the author on the Colorado Plateau in the southwestern United States and in the Pre-Saharan depression in southern Morocco (SCHMIDT 1987a, 1987b). Both regions are intermontane areas with altitudes above 1000 m in the Pre-Saharan depression and above 1500 m on the Colorado Plateau. Mean annual precipitation amounts to 150–250 mm in the central parts of the Colorado Plateau and to 100–175 mm in the Pre-Saharan depression between the High Atlas and the Anti-Atlas. The southwestern United States as well as southern Morocco experienced a wetter climate during the last glacial (WELLS et al. 1982, ROGNON 1987).

A cuesta scarp is a compound form generally consisting of an upper vertical cliff in a resistant caprock and a moderately inclined lower slope in the underlying softer rock. A crucial question is whether the alternating climatic conditions have had a significant bearing on the processes which, through erosion and deposition, shaped the profile of the cuesta scarps. This question will be discussed in comparing similar slope forms of the Colorado Plateau and southern Morocco.

2 Talus and pediment flatirons — description

Talus and pediment flatirons on scarp slopes are the results of transport, depositional and erosional processes. Talus flatirons and related landforms have been described from the southwestern United States (KOONS 1955, BLUME & BARTH 1972, SCHIPULL 1980), from the Saharan countries (ERGENZINGER 1972, GRUNERT 1983), from Morocco (JOLY 1962, DONGUS 1980, SCHMIDT 1986), from Saudi-Arabia (BARTH 1976), from Syria (SAKAGUCHI 1986) and from Israel (GERSON 1982, GERSON & GROSSMAN 1987).

Flatirons have a triangular to trapezoidal shape with their top directed towards the cuesta scarp (fig.1, photo 1). They are detached from the active scarp slope, by which they were formerly fed with debris. The inner slope, which is inclined towards the scarp, consists of

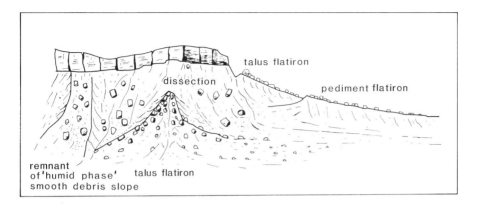

Fig. 1: *The cyclic model of talus and pediment flatiron development.*
A smooth concave debris slope is formed in the more humid phase (pluvial), it becomes dissected and steepened in the more arid phase (interpluvial). Parts of the former slope are detached from the active scarp by dissection and scarp retreat, and talus and pediment flatirons originate.

Photo 1: *Cuesta scarp with caprock of Mio-Pliocene conglomerates in the Pre-Saharan depression in SE Morocco.*
The last pluvial smooth debris slope covers wide areas in the scarp foreland. In the foreground a talus flatiron has been formed with its inner slope (light grey) directed towards a scarp to the left. Note the thin debris cover on the backslope. In the background another talus flatiron has been detached from the scarp slope by dissection and scarp retreat.

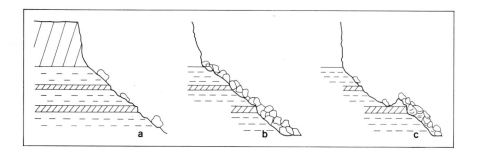

Fig. 2: *The non-cyclic model of talus flatiron development.*
The lower slope is steepened (40°) and dissected by rills, until a critical state of caprock destabilization is reached (a), a rockfall is triggered, which covers the complete lower slope with a talus cone (b), the cone is undermined by rills at its flanks and at its apex, it becomes detached from the active scarp slope, and a talus flatiron originates (c).

bedrock material. The outer slope, which is inclined towards the scarp foreland with slope angles from 20 to 2°, is protected by a relatively thin debris cover. The talus consists of caprock material, which is commonly cemented by calcium carbonate, and attains an average thickness of 2 m (cf. SCHMIDT 1987b); thicknesses of more than 5 m are rare (KOONS 1955, GERSON 1982). In the proximal parts, however, the debris may contain talus blocks with diameters of more than 2 m. Talus flatirons are located on the scarp slope and pediment flatirons in the transition zone between scarp and foreland. In the distal parts pediment flatirons may find their continuation in terraces. Both talus and pediment flatirons are remnants of formerly active continuous slope systems.

3 Models of talus and pediment flatiron development

Different views have been offered concerning the processes and stages responsible for the formation of talus and pediment flatirons. These views can be categorized into two principal models. The first one states that the flatirons are formed within a climatically undifferentiated phase (**non-cyclic model**), the second one states that flatirons are the product of processes of climatically distinct phases (**cyclic model**). It is noteworthy that those who have worked in the United States support the non-cyclic model, and those who have worked in Northern Africa or in the Middle East support the cyclic model.

The non-cyclic model conceives the following succession of events (KOONS 1955, SCHIPULL 1980, SCHMIDT 1987a) (fig.2). The caprock is undermined by slope gullies and/or steepening of the lower slope. Rockfalls are triggered and cover the lower slopes with talus cones. These are undermined by slope rills at their flanks. At the same time their apex is detached from the scarp slope by small tributaries, which run perpendicular to the main slope rills. By this process the talus cone becomes isolated from the scarp and eventually a talus flatiron comes into existence. The scarp slope is steepened again and dis-

Photo 2: *Cuesta scarp with caprock of Cenomanian/Turonian limestone in the Pre-Saharan depression, SE Morocco.*
A talus flatiron is located on the scarp slope in the background and a pediment flatiron with a gently inclined backslope in the foreground. The photo was taken close to the profile of fig.3.

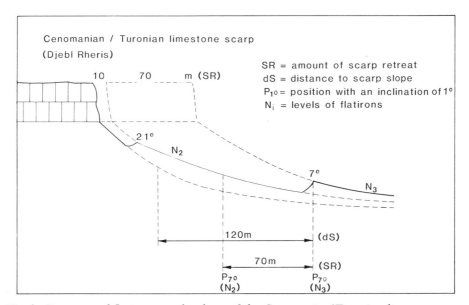

Fig. 3: *Sequence of flatirons on the slope of the Cenomanian/Turonian limestone scarp of the Djebl Rheris, SE Morocco.*
N_2 and N_3 are levels of talus (pediment) flatirons. N_1 is the youngest (Holocene) terrace level. The amount of scarp retreat can be obtained by measuring the distance from the top of the older flatiron (P_7o; N_3) to the point of equal inclination on the next younger flatiron backslope (P_7o; N_2). The amount of retreat in the example is 70 m. Measuring the distance between the tops of the flatirons or the distance to the scarp slope will overestimate the amount of retreat. As the cycle from Riss to Würm lasted about 10^5 years, a rate of retreat of 0.7m/1000 years is obtained. Since the last pluvial the scarp has retreated about 10 m. The average rate of retreat of the limestone scarps in Southern Morocco is about 0.5 m/1000 years (SCHMIDT 1987b).

sected by rills, until a critical state of caprock destabilization is reached, which triggers a new rockfall. No climatic change is needed for this succession of events. The successive stages of talus flatiron formation from fresh talus cones to complete separation can be observed in the field in close spatial connection.

In the cyclic model talus and pediment flatirons are landforms which represent climatically dependent alternations of processes on scarp slopes (cf. GERSON 1982, SCHMIDT 1986, 1987b). It is only in this model that pediment flatirons are also considered. The outer slopes or backslopes of the flatirons are the product of slopewash and debris throughput with temporary deposition on smooth undissected slopes during humid phases. There was some gravitational activity in the proximal parts. The smooth debris slopes are dissected in the more arid periods, bedrock is exposed in the rills; gullying and rock collapse are the dominant processes. Parts of the former slope are detached from the active scarp by dissection and scarp retreat, and talus and pediment flatirons originate (fig.1, photo 1 & 2).

Hillslopes in drylands react in a far less sensitive way to changing climatic conditions than do flood plains or river beds, which are also affected by slight environmental changes of short duration (cf. COOKE & REEVES 1976, EULER et al. 1979). Slope systems have relatively long relaxation times, and GERSON (1982) gives a tentative estimate of 25,000 to 50,000 years for a smooth debris slope to develop after dissection. In front of many cuesta scarps, different generations of talus relics are found; each flatiron's outer slope representing the latest stage of debris slope activity in a pluvial phase (fig.3, photo 2). Thus the distance of a sequence of flatirons is an indicator for the amount of scarp retreat in a complete interpluvial/pluvial cycle (GERSON 1982, SCHMIDT 1986, 1987b).

4 Application of the models to the southwestern United States and to southern Morocco

On the Colorado Plateau there is a great variety of different caprocks which, in conjunction with the underlying softer rocks, form different compound slope types on the cuesta scarps. Caprocks involved are sandstones with different degrees of consolidation and different cementing materials, conglomerates and only few limestones. The influence of the soft rocks on the morphology of the lower slope has been described in greater detail in an earlier article (SCHMIDT 1987a). The following four slope types are distinguished:

- (I) concave lower slopes (35°), turning into pediments at the scarp foot, in impermeable and homogeneous shales,

- (II) convex slopes with lower inclinations (30°) in swelling bentonitic clays, influenced by piping processes and high-magnitude landsliding events,

- (III) slopes composed of segments of different inclination in heterogeneous lower slope rocks with intercalations of more resistant beds, the resistant beds impeding deep dissection, in the overall profile the slopes are straight with inclinations of about 40°.

- (IV) vertical cliffs with talus cones at their base in evenly bedded, mostly gypsiferous fine sandstones, siltstones and shales with vertical joints.

Most talus flatirons on the Colorado Plateau are found on the straight slopes in the heterogeneous lower slope-forming rocks. The high slope inclination (40°) and the dense, though not deep dissection further the generation of rockfalls with talus cone formation. Talus flatiron development on this slope type is subject to the succession of events described for the non-cyclic model, i.e. talus cone production and separation within one climatic period. Pediment flatirons, which are transitional to foreland levels, are not found on these scarps.

On the Colorado Plateau talus and pediment flatirons of the cyclic type are found on the concave lower slopes in the thick homogeneous shales and marls. Not being very numerous on the whole, they have been observed and described in the Mesa Verde area (BLUME & BARTH 1972), where they are developed on Mancos Shale slopes, and in the Grand Canyon area, where they are found on Hermit Shale slopes (KOONS 1955) below the resistant Kaibab limestone.

In southern Morocco the principal caprocks are Palaeozoic quartzites, Mesozoic and Palaeogene limestones and Mio-Pliocene conglomerates (SCHMIDT 1987b), all of which are underlain by soft redbeds generally consisting of marls and poorly consolidated siltstones and sandstones. The characteristic lower slope profile in the soft rocks is concave. Talus **and** pediment flatirons have developed independently of the different caprocks involved on most cuesta scarps (photo 1 & 2). In southern Morocco slope dissection is at present an ubiquitous process and no smooth debris slopes are formed. Here, flatiron formation is clearly the result of a sequence of different, climatically-controlled slope processes, which was difficult to realize given the contrasting evidence from the Colorado Plateau, where the majority of flatirons are non-cyclic.

5 Prerequisites for the different modes of talus and pediment flatiron development

Cuesta scarps with non-cyclic talus flatirons must be susceptible to high magnitude rockfall events, which cover the complete lower slope with a talus cone. This requires a thickness ratio between soft rock and caprock not much greater than unity. On the other hand, lower slope inclination must be high enough to destabilize greater portions of the caprock and the slope must not be deeply dissected, otherwise the talus is trapped in the slope rills and no talus cones can develop. All these conditions are met in the straight slope type of the Colorado Plateau, where the lower slope rocks are overlain by thick sandstones. Rockfall events also sporadically occur on some Moroccan cuesta scarps, and here talus flatiron development provides the only exception from cyclic processes in southern Morocco (cf. SCHMIDT 1986).

Though both the southwestern United States and southern Morocco have experienced climatic change in recent earth history, from moister to drier conditions, no analogous morphological change has been observed on the cuesta scarps. There must be other controlling fac-

(1)	thick lower slope rocks of homogeneous resistance showing a concave profile
(2)	resistant caprocks, carbonate cement
(3)	thickness ratio between 1 and 3
(4)	no complex scarps
(5)	no subsequent valleys at scarp foot or deeply incised canyons in scarp foreland

Tab. 1: *Conditions favourable for cyclic flatiron development.*

Photo 3: *Cuesta scarp near Hanksville, Utah, with a thickness ratio between soft rock and caprock of more than 5.*
The caprock is a less resistant intra-formational member of the Mancos Shale (Emery Sandstone) on the highly erodible Blue Gate Shale. The caprock is not thick enough to cover the complete lower slope with a talus veneer.

tors of lithological and structural nature, which favour or impede the cyclic development of talus and pediment flatirons, which is predominant in southern Morocco. The following conditions should be met (tab.1):

- (1) A smooth concave debris-flow controlled slope must have been formed in the more humid phase before its dissection may begin in the more arid period (fig.1). Such slope formation is only possible in thick strata of relatively uniform resistance like the redbeds in the lower slopes in southern Morocco.

- (2) The outer slopes of the talus and pediment flatirons can only survive periods of cyclic duration, if the talus on the backslopes is very resistant to weathering and erosion. Resistant talus can only be supplied by very resistant caprocks. Limestones, quartzites, conglomerates, but also well-cemented sandstones yield highly protective talus covers. The protective effect is en-

Photo 4: *Valley-side cuesta scarp near Moab, Utah.*
The caprock consists of sandstones of the Glen Canyon Group; the lower slope lies in the Chinle Formation. The Colorado River flows at the foot of the scarp.

	Total number of cuesta scarps investigated		105
(1)	Number of scarps with lower slope rocks which form	convex lower slopes straight lower slopes vertical lower slopes	18 38 19
(2)	Number of scarps in less resistant sandstones (e.g. Navajo Sandstone)		27
(3)	Number of scarps with thickness ratios	greater than 5 smaller than 1	26 4
(4)	Number of complex scarps		41
(5)	Number of scarps with subsequent valley at scarp foot		10
		Total	183
	From the total number (183), one can conclude that most scarps are influenced by one or more lithological and structural constraints, which impede cyclic flatiron development		

Tab. 2: *Attributes of cuesta scarps on the Colorado Plateau, which are unfavourable for cyclic talus and pediment flatiron formation.*

hanced, when the talus becomes cemented by calcium carbonate. Limestone caprocks or carbonate-cemented conglomerates and sandstones provide the cementing material needed for caliche (calcrete) formation. In southern Morocco the caprocks either guarantee the supply of carbonate cement (limestones, Mio-Pliocene conglomerates) and/or are very resistant (limestones, quartzites).

- (3) Another factor, other than soft rock and caprock lithology, is the thickness ratio between soft rock and caprock. When the caprocks are relatively thin with thickness ratios of more than 5, no flatirons have developed, because the caprock is not thick enough, especially in not too resistant caprocks, to supply the lower slope with a continuous talus cover (photo 3). On the other hand, when the thickness ratio is less than unity, talus production may exceed talus transport and removal, throughput equilibrium is disturbed, thick talus accumulation occurs, and flatiron formation is not possible.

- (4) On complex cuesta scarps where more than one resistant caprock and underlying soft rock is involved in scarp formation, talus flatiron development cannot occur.

- (5) Subsequent rivers or washes at the scarp foot (cf. GERSON & GROSSMAN 1987) or deeply incised canyons in the immediate scarp foreland (photo 4) make the formation of concave basal slopes and pediment levels, and thus the development of cyclic flatirons, impossible.

In southern Morocco conditions are very favourable for the cyclic development of talus and pediment flatirons (tab.1), on the Colorado Plateau, however, only some of the lower slope rocks allow the generation of concave profiles — convex, straight and vertical slopes frequently occur. A number of caprocks consist of less resistant sandstones, and the thickness ratio in many cases exceeds a value of 5, and is often less than unity (tab.2). Moreover, a great number of scarps are complex and/or are accompanied by subsequent valleys (tab.2), and deeply incised canyons are a major feature of the Colorado Plateau.

6 Conclusion

Environmental change may have a geomorphological effect on slope processes, but this is not necessarily the case. In one system flatiron surfaces may be associated with pluvials, but in others lithological and structural constraints do not allow a climatic change to become morphodynamically effective and cause a succession of climato-genetic landforms (BÜDEL's (1982) sense). It will depend on the height of the threshold that has to be overcome to shift a slope system from one state of equilibrium to another. Where thresholds are low and conditions are favourable, as in southern Morocco, system change may occur even if there is no fundamental change of climate. In this case climato-genetic views are given preference. Where thresholds are high and lithological and structural controls strong, and where non-cyclic sequences are observed as on the Colorado Plateau, structurally-dominated views are given priority.

It is important to note that

- the absence of climato-cyclic landforms does not necessarily indicate the absence of climatic change; it only means that changes may not have exceeded a given threshold,

- the presence of landform generations caused by climatic change does not mean that climate is the prime controlling factor of relief-forming processes, nor that the same successions are regularly developed in all regions having the same climatic history.

Climate is only one controlling factor among others, and the contrasting views supported in the regions under investigation are due to the regionally-different resistances to change.

References

BARTH, H.K. (1976): Pedimentgenerationen und Reliefentwicklung im Schichtstufenland Saudi-Arabiens. Z. Geomorph. N.F. Suppl. Bd. **24**, 111–119.

BLUME, H. & BARTH, H.K. (1972): Rampenstufen und Schuttrampen als Abtragungsformen in ariden Schichtstufenlandschaften. Erdkunde **26**, 108–116.

BÜDEL, J. (1982): Climatic Geomorphology. Princeton University Press, Princeton.

COOKE, R.U. & REEVES, R.W. (1976): Arroyos and Environmental Change in the American South-West. Clarendon Press, Oxford.

DONGUS, H. (1980): Rampenstufen und Fußflächenrampen. Tübinger Geographische Studien **80**, 73–78.

EULER, R.C., GUMERMAN, G.C., KARLSTROM, T.N.V., DEAN, J.S. & HEVLY, R.H. (1979): The Colorado Plateaus: cultural dynamics and palaeoenvironment. Science **205**, 1089–1101.

ERGENZINGER, P. (1972): Reliefentwicklung an der Schichtstufe des Massif d'Abo (Nordwest-Tibesti). Z. Geomorph. N.F. Suppl. Bd. **15**, 93–112.

GERSON, R. (1982): Talus relicts in deserts: a key to major climatic fluctuations. Israel Journal of Earth Sciences **31**, 123–132.

GERSON, R. & GROSSMAN, S. (1987): Geomorphic activity on escarpments and associated fluvial systems in hot deserts as an indicator of environmental regimes and cyclic climatic changes. In: RAMPINO, R., SANDERS, J.E., NEWMAN, W.S. & KÖNIGSSON, L.K. (Eds.), Climate, History, Periodicity, Predictability. Van Nostrand Reinhold, New York, 300–322.

GRUNERT, J. (1983): Geomorphologie der Schichtstufen am Westrand des Murzuk-Beckens (Zentrale Sahara). Relief, Böden, Paläoklima, **2**, 1–269.

JOLY, F. (1962): Etudes sur le relief du Sud-Est Marocain. Trav. Inst. Sc. Chér., Sér. Géol. et Géogr. Phys. (Rabat), **10**, 1–578.

KOONS, D. (1955): Cliff retreat in the southwestern United States. American Journal of Science, **253**, 44–52.

ROGNON, P. (1987): Late Quaternary climatic reconstruction for the Maghreb (North Africa). Palaeogeography, Palaeoclimatology, Palaeoecology, **58**, 11–34.

SAKAGUCHI, Y. (1986): Pediment — A glacial cycle topography? Genesis of the pediments in central Syria. Bulletin of the Department of Geography, University of Tokyo, **18**, 1–19.

SCHIPULL, K. (1980): Die Cedar Mesa-Schichtstufe auf dem Colorado Plateau — ein Beispiel für die Morphodynamik arider Schichtstufen. Z. Geomorph. N.F. **24**, 318–331.

SCHMIDT, K.-H. (1986): Strukturbestimmte Relieftypen und Tektonik im Grenzbereich zwischen Hohem Atlas und Anti-Atlas. Berliner Geow. Abh., **A 66**, 503–514.

SCHMIDT, K.-H. (1987a): Factors influencing structural landform dynamics on the Colorado Plateau — about the necessity of calibrating theoretical models by empirical data. CATENA SUPPLEMENT **10**, 51–66.

SCHMIDT, K.-H. (1987b): Das Schichtstufenrelief der präsaharischen Senke, Süd-Marokko. Z. Geomorph. N.F. Suppl. Bd. **66**, 23–36.

WELLS, S.G., BULLARD, T.F. & SMITH, L.N. (1982): Origin and evolution of deserts in the Basin and Range and the Colorado Plateau provinces of western North America. Striae, **17**, 101–111.

Address of author:
Karl-Heinz Schmidt
Institut für Physische Geographie
Geomorphologisches Labor der FU
Altensteinstr. 19
D-1000 Berlin 33
West Germany

FLUVIAL SYSTEMS OF THE KALAHARI
A REVIEW

P. Shaw, Gaborone

Summary

Although surface drainage is absent from much of the Kalahari, three distinct types of fluvial systems are apparent in the northern and eastern parts of Botswana, and in the adjacent northern Cape Province. The Okavango Delta, with its associated outflows and palaeo-lake basins, dominates the northern areas. Sedimentation within the Okavango graben is dominantly of fine sand, with duricrust formation from the solute load in the distal sector of the Delta. East of the Kalahari-Zimbabwe divide aggressive seasonal streams drain to the Limpopo, often superimposed on outliers of Kalahari Sand. Many of these rivers are of the "Sand River" type and form an important groundwater resource. West of the Kalahari-Zimbabwe divide and straddling the "Bakalahari Schwelle", surface drainage is represented by pans and massive fossil valleys such as the Molopo and Okwa-Mmone systems, which rarely contain surface water at the present time. Recent research has indicated that these valleys frequently occupy pre-Kalahari lineations and are partly attributed to groundwater flow. The evolution of these drainage forms is examined within the context of tectonics and climatic change.

1 Introduction

The massive Kalahari erg covers some 2.5 m km^2, and extends from Zaire to South Africa. Two thirds of Botswana (area 582,000 km^2) is covered by sands of the Kalahari Series, defined by PASSARGE (1904) as the Middle Kalahari, extending from the Zambezi southwards to approximately 23°S, and the Southern Kalahari, extending into the northern Cape Province. The climate in this region is semi-arid (Koppen BSh), with precipitation varying from <250 mm in the southwest to >650 mm in the Zambezi Valley.

Although the population of Botswana is presently estimated to be only slightly over 1 million, the persistent edaphic drought and urgent demand for water for industrial and urban growth places heavy emphasis on both surface and groundwater resources, the latter supplying some 80% of the total demand at present.

Fluvial channel systems are absent from much of the central and southern parts of the country, where surface drainage is represented by hundreds of pans (LANCASTER 1978) which may contain water after seasonal rains. Elsewhere the fluvial systems of the country fall into three distinct types, each with

ISSN 0722-0723
ISBN 3-923381-17-4
©1989 by CATENA VERLAG,
D-3302 Cremlingen-Destedt, W. Germany
3-923381-17-4/89/5011851/US$ 2.00 + 0.25

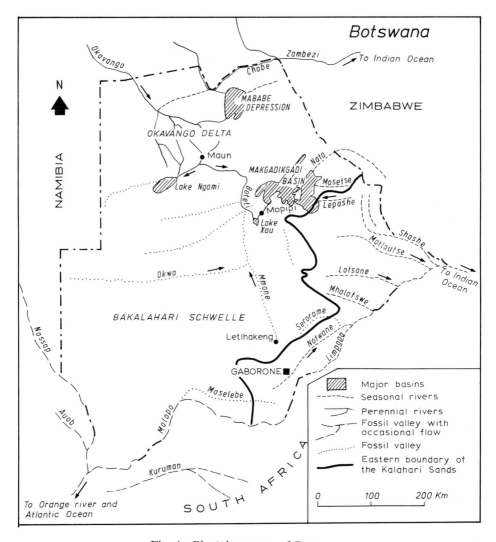

Fig. 1: *Fluvial systems of Botswana.*

characteristic regimes and geomorphological processes (fig.1). This paper summarizes the characteristics of these fluvial systems, and accounts for their evolution within the context of tectonism and climatic change.

2 The Okavango and perennial rivers of the north

Nothern Botswana is dominated by the three linked networks of the Okavango, Chobe and Zambezi Rivers, with the last two forming, in part, international boundaries. The Okavango flows into Botswana at Shakawe and spreads out to form the Okavango Delta, a unique alluvial fan with sediments covering 22,000 km². The present flooded area varies from 6,000 to 13,000 km² depending on prevailing conditions (UNDP/FAO 1977). At present links to the Chobe and Zambezi Rivers are tenuous, though they have been more firmly established during the late Quaternary (SHAW 1985a, SHAW & THOMAS, 1988).

The annual budget of the Okavango amounts to 10.5×10^9 m³ year^{-1} inflow from the Angolan Highlands, together with 5×10^9 m³ year^{-1} precipitation over the Delta area, with the former arriving as a series of flood pulses from April onwards, and the precipitation occuring largely between October and March. Some 93% of the total water is lost through evapotranspiration; the remainder leaving as groundwater or as outflow towards the Chobe River, or to the Ngami, Mababe and Makgadikgadi Basins beyond the Delta periphery. These basins have formed important components of a complex Quaternary palaeolake system whose form and function have been detailed elsewhere (COOKE 1980, COOKE & VERSTAPPEN 1984, SHAW 1985a, SHAW & COOKE 1986). At present the Ngami Basin contains an ephemeral lake fed from the Kunyere and Lake River outflows, whilst the Boteti flows anually towards Lake Xau and the Mopipi swamps on the western side of the Makgadikgadi, where it forms an important water resource for the Orapa diamond mine. All three basins ar fault controlled; the Mababe and Ngami Basins lie against the Thamalakane Fault, which terminates the Delta and is an extension of the East African Fault system. Sediment thickness attains 600 m along this fault, representing a prolonged period of subsidence in this tectonically active zone (REEVES 1972).

Flow within the swamps of the Delta is erratic, and the routes taken by the water vary from year to year in response to sedimentation, minor tectonism, vegetation blockage and the activities of aquatic animals, particularly the hippopotamus. As pressure to develop the Okavango as a water resource for the 21st century has mounted (UNDP/FAO 1977, SMEC 1986), the understanding and modelling of these subtle changes has become increasingly important (HUTTON & DINCER 1979), and must include estimation of the sediment budget.

Sedimentation within the Okavango swamps may be considered in terms of solute, flotation and suspended/bedload components. The latter has been estimated at 0.6 million tons (400,000 m³) per year (UNDP/FAO 1977) with the mean grain size of 0.2–0.4 mm in the upper Delta, grading down to 0.2 mm in the distal sector. The inadequacy of the sampling programme to date suggests that the figure is too low, and may, in fact, reach 2 million tons (1.3 million m³) per year (SMEC 1986). Deposition of bedload is certainly not uniform throughout the swamps. Recently McCARTHY et al. (1986a) have proposed a model of aggradation and vegetation stabilisation which would initiate a cycle of channel growth and decay on a decadal scale, and would account for the raising of channels above swamp level. Vegetation undoubtedly plays an important role, both as a stabilising influence and as a flotation load, as the diminution in the size of the 19th century Lake Ngami illustrates (SHAW 1985b).

The role of the solutes is less well known. 18-Oxygen studies of the Delta

(DINCER et al. 1978) have indicated increasing evaporation downstream, whilst groundwater analysis (HUTCHINS et al. 1978) shows a complex pattern of low TDS (<1 gm l^{-1}) calcium-dominated water near the channels, and older water (TDS 1–5 gm l^{-1}), rich in chlorides and sulphates, in parts of the lower delta. Altogether salt concentrations are surprisingly low, given the prevailing evaporation rates in the Delta and the formation of massive salt deposits in the adjacent Makgadikgadi. Recent studies have indicated that one reason for this may be precipitation of salts, particularly calcite, trona and thermonatrate, on the shorelines of islands within the swamp complex (McCARTHY et al. 1986b).

SUMMERFIELD (1982) has also indicated that silica concentrations in Okavango water are high (mean 55 ppm for 24 samples) and increase downstream. Recent investigations in the vicinity of the Thamalakane and upper Boteti Rivers (SMEC 1987) indicate laminated crypto-crystalline silcrete forming to the limits of drilling at 20 m. Contemporary formation of silcrete is rare, and studies of this phenomena are being undertaken at present.

3 The seasonal rivers of eastern Botswana

Eastern Botswana has a more conventional drainage network developed, for the most part, to the east of the Kalahari Sand boundary, and is thus marginal to the main Kalahari drainage. Th network rises along the Kalahari-Limpopo watershed, roughly coincident with the alignment of the Kalahari-Zimbabwe Rise. the influence of Tertiary uplift along this axis is apparent from the strong development of the left bank tributaries of the Limpopo network, which rise on the fringes of the Kalahari Sands, or in isolated hill outcrops, before flowing northeast across the African planation surface. In northeast Botswana seasonal rivers such as the Lepashe, Mosetlhe and Nata also drain the hardveld westwards to Sua Pan.

The rivers vary considerably in character. Beyond the rock-controlled hill segments two basic tendencies emerge. Rivers such as the Limpopo and its tributaries in the southwestern part of the country, such as the Notwane, flow between a series of rock basin, confined by a flood plain of alluvial clays. The sand load of these rivers is very small and confined to sandbanks within the channel. A characteristic of these rivers is the presence of extensive terrace sequences upto 20 m above the present channel, frequently composed of large pebbles and cobbles originally derived from the Waterberg Series conglomerates. WAYLAND (1950) suggested that the present Notwane River has re-excavated earlier infill derived from a different climatic regime, but the hypothesis has never been proved.

The rivers of northeastern Botswana, particularly the Shashe, Tati and Moutloutse, are "sand rivers", with wide, low-gradient beds (0.15–0.20%) of sand and gravel up to 15 m in depth, over Pre-Cambrian bedrock. The larger rivers of this type have sand beds throughout their length, whilst others, notably tributaries of the upper Notwane, such as the Metsemotlhaba, have sand sections only in their upper catchments.

NORD (1985) believes the sand to originate from the erosion of granite in the valleys under humid conditions, suggesting that the present configuration is a fossil form which the river is unable to shift, though there is no reason to assume that equilibrium has not been attained under the present regime. The juxtaposition of sand and rockbed rivers is common in southern Arica, and was noted in the early years of European expansion (e.g. SELOUS 1893).

The rivers of both types are dry throughout much of the year, but are prone to flash flood events during the summer months, exacerbated by the increasing incidence of land degradation, particularly in the Central District (VIAK 1984). NORD (1985) estimates the extreme dry period for these rivers at 10 months, with no-flow years a rarity; the Mahalapshwe River, for example, has failed to flow only once in the past 100 years. In good rains several flood events can be expected.

The sand rivers now supply approximately 33% of Botswana's groundwater and have the advantage of being a perenially recharged source. Despite major investigations into their aquifer potential (WIKNER 1980, NORD 1985), and the practical problems of dam siltation, no research has been conducted into the sediment budgets of the eastern rivers, though implementation of research programmes on suspended and bedload components is in hand.

4 The fossil valleys of the Southern Kalahari

A prominent landform in the Southern Kalahari is the Mokgacha, a large dry valley, sometimes one or more km in width, characterized by a flat floor, steep duricrusted sides and abrupt valley heads. Two principal networks of mekgacha occur. The first is the Molopo River, with its tributaries in South Africa, the Auob, Nossop and Kuruman, which drain the southern Kalahari to the Atlantic. The second is the Okwa-Mmone system, which provides surface drainage from both eastern Namibia and eastern Botswana towards the Makgadikgadi Basin. The broad interfluve of the southern Kalahari between these networks has been termed the "Bakalahari Schwelle" (PASSARGE 1904) and forms an area of concentrated pan development. Smaller mekgacha occur in the Middle Kalahari to the west of the Makgadikgadi and Ngami Basins, whilst the unusual Serorome valley in eastern Botswana, initiated on the western flank of the Kalahari watershed, curves through 180 degrees to the Zoetfontein Fault Block and Limpopo river.

These mekgacha landforms are usually attributed to fluvial erosion during periods of wetter climate. The Molopo system, at least, bears some resemblance to a river channel, and responds to short-term precipitation events. Sporadic floods have been recorded in the Nossop in 1806, 1963 and January 1987, in the Kuruman in 1894 and 1918, and in all four rivers of the Molopo network in 1934 (CLEMENT 1967). Inevitably these floods occur in response to rainfall over the upper catchment and are short-lived, with the water being rapidly absorbed into the river bed. They may, however, be of considerable magnitude; the 1934 Nossop flood was estimated by the newspaper of the time as being 450 feet wide, travelling at 6 mph (CLEMENT 1967).

Less spectacular, but probably more important in the longterm evolution of these landforms, is the role of groundwater. PEEL (1941) suggested that fossil valleys in tectonically stable arid regions could be formed by spring sapping and groundwater erosion. This is most evident in the Kuruman valley, which is fed by the famous eye of Kuruman, a dolomite spring which has yielded a constant 750 m^3 hour^{-1} since at least 1820. Some of the water is used for agriculture, the remainder disappears into the river bed a few km downstream.

The mekgacha of the Okwa-Mmone system, however, have not carried surface flow during the period of historical records, although some parts have sustained higher water tables in the 19th century, now diminished by increased groundwater extraction for the cattle industry (CAMPBELL & CHILD 1971). Large sections of the valleys, particularly in the central Kalahari, are now obscured by an infill of Kalahari Sand.

Investigation of the Lethlakeng Valleys at the head of the Mmone in eastern Botswana (SHAW & DE VRIES 1988) has shown that these impressive gorges are formed by deep weathering and groundwater erosion along pre-Kalahari lineaments, and have become incised since the uplift of the Zimbabwe-Kalahari Axis in the Tertiary. The complex duricrust suite encountered in the valleys is formed from Karoo sedimentary rocks, pre-existing calcretes and spring deposits, with variations dependant upon the nature of the parent matrial and the hydraulic gradient.

The role of deep weathering is also becoming increasingly apparent in the formation of pans in the Kalahari. Recent geochemical and geophysical studies of pans (BUTTERWORTH et al. 1982, FARR et al. 1982) have indicated alteration and duricrusting of both Kalahari and Karoo strata to depths exceeding 27 m, whilst ARAD's (1984) study of the Molopo region has suggested aquifer recharge through dry channels, wth many pans aligned to pre-Kalahari drainage. The pattern now emerging from the Southern Kalahari is one of long-term drainage evolution by groundwater processes, involving solute transport and duricrust formation, interspersed with periods of sporadic fluvial and aeolian activity. These geomorphic processes are probably more important in sediment distribution than in actual erosion, given the low gradients and resistance of the duricrust outcrops.

5 Conclusion

The Middle and Southern Kalahari, occupying Botswana and the northern Cape, has three types of hydrological regime, characterized by perennial, seasonal and groundwater flows respectively. The drainage pattrn of the Southern Kalahari and adjacent parts of eastern Botswana, a tectonically stable area, has been evolving since the Tertiary, whilst the Okavango system in its present form is a relatively recent feature, liable to further change with or without human interference. All three drainage types have been strongly influenced by past climatic conditions, particularly those involving greater precipitation, and are now moving towards new equilibrium forms.

At present little is known of the sediment budgets of any of the drainage systems, though work so far suggests that the solute component may be more important than previously thought. Sediment data collection has now been initiated in a number of sites, and promises a valuable insight into the form and evolution of these networks in years to come.

References

ARAD, A. (1984): Relationship of salinity of groundwater to recharge in the southern Kalahari Desert. Journal of Hydrology, 71, 225–238.

BUTTERWORTH, J.S. (1982): The chemistry of Mogatse Pan - Kgalagadi District. Botswana Geol. Surv. Dept. Report JSB/124/82.

CAMPBELL, A.C. & CHILD, G. (1971): The impact of man on the environment of Botswana. Botswana Notes & Records, 3, 91–109.

CLEMENT, A.J. (1967): The Kalahari and its Lost City. Longmans, Cape Town.

COOKE, H.J. (1980): Landform evolution in the context of climatic change and neo-tectonism in the Middle Kalahari of north central Botswana. Transactions of the Institute of British Geography, NS 5, 80–99.

COOKE, H.J. & VERSTAPPEN, B.Th. (1984): The landforms of the western Makgadikgadi Basin in northern Botswana, with a consideration of the chronology of the evolution of Lake Palaeo-Makgadikgadi. Zeitschrift für Geomorphologie, NF 28, 1–19.

DINCER, T., HUTTON, L. & KHUPE, B. (1978): Study, using stable isotopes, of flow distribution, surface-groundwater relations and evapotranspiration in the Okavango Swamp, Botswana. Isotope Hydrology 1978, Int. Atomic Energy Auth., Austria.

FARR, J., PEART, R., NELISSE, C. & BUTTERWORTH, J. (1982): Two Kalahari Pans: a study of their morphometry and evolution. Botswana Geol. Surv. Dept. Report GS10/10, Gaborone.

HUTCHINS, D., HUTTON, L., HUTTON, S., JONES, C. & LEONHERT, E. (1976): A summary of the geology, seismicity, geomorphology and hydrogeology of the Okavango Delta. Botswana Geol. Surv. Dept. Bull, 7, Gaborone.

HUTTON, L. & DINCER, T. (1979): Using satellite imagery to study the Okavango Swamp, Botswana. In: Satellite Hydrology. DEUTSCH, M., WIESNET, D. & RANGO, A. (Eds.), American Water Resource Association.

LANCASTER, N. (1978): The pans of the southern Kalahari, Botswana. Geographical Journal, 144.

McCARTHY, T., ELLERY, W., ROGERS, K., CAIRNCROSS, B, & ELLERY, K. (1986a): The role of sedimentation and plant growth in changing flow patterns in the Okavango Delta. South African Journal of Science, 82, 579–584.

McCARTHY, T., McIVER, J. & CAIRNCROSS, B. (1986b): Carbonate accumulation on islands in the Okavango Delta. South African Journal of Science, 82, 588–591.

NORD, M. (1985): Sand rivers of Botswana: Phase II. Dept. Water Affairs/SIDA, Gaborone.

PASSARGE, S. (1904): Die Kalahari. Dietrich Riemer, Berlin.

PEEL, R.F. (1941): Denudational landforms of the central Libyan Desert. Journal of Geomorphology, 4, 3–23.

REEVES, C.V. (1972): Evidence of rifting in the Kalahari. Nature Physics and Science, 237, 96.

SELOUS, F.C. (1893): Travel and Adventure in South East Africa. Rowland Ward, London.

SHAW, P.A. (1985a): Late Quaternary landforms and environmental change in northwest Botswana: the evidence of Lake Ngami and the Mababe Depression. Transactions of the Institute of British Geographers, NS 10, 333–346.

SHAW, P.A. (1985b): The desication of Lake Ngami: a historical perspective. Geographical Journal, 151, 318–326.

SHAW, P.A. & COOKE, H.J. (1986): Geomorphic evidence for the late Quaternary palaeoclimates of the middle Kalahari of northern Botswana. CATENA, 13, 349–359.

SHAW, P.A. & DE VRIES, J.J. (1988): Duricrust, groundwater and valley development in the Kalahari of southeast Botswana. Journal of Arid Enviroments, 14, 245–254.

SHAW, P.A. & THOMAS, D.S.G. (1988): Lake Caprivi: a late Quaternary link between the Zambezi and Middle Kalahari drainage systems. Zeitschrift für Geomorphologie, NF 32, 329–337.

SMEC (SNOWY MOUNTAINS ENGINEERING CORPORATION) (1986): Southern Okavango Integrated Water Development, Phase I, Inception Report. Dept. Water Affairs/SMEC, Gaborone.

SMEC (SNOWY MOUNTAINS ENGINEERING CORPORATION) (1987): Southern Okavango Integrated Water Development, Phase I, Final Report, Technical Study, Volume 1. Dept. Water Affairs/SMEC, Gaborone.

SUMMERFIELD, M.A. (1982): Distribution, nature and probable genesis of silcrete in arid and semi-arid southern Africa. In: Aridic soils and geomorphic processes. (Ed. D.H. YAALON) CATENA SUPPLEMENT 1, 37–65.

UNDP/FAO (1977): Investigation of the Okavango Delta as a primary water resource for Botswana. DP/BOT/71/506 Tech. Report (3 Vols). UNDP, Gaborone.

VIAK, (1984): Eastern Botswana Regional Water Study. Dept. Water Affairs/VIAK (2 Vols). Gaborone.

WAYLAND, E. (1950): Unpublished papers. National Museum of Botswana, Gaborone.

WIKNER, T. (1980): Sand Rivers of Botswana — Phase I. Dept. Water Affairs/SIDA. (2 Vols). Gaborone.

Address of author:
P. Shaw
Department of Environmental Science
University of Botswana
P/Bag 0022 Gaborone
Botswana

INFLUENCE OF EOLIAN SEDIMENTATION ON SOIL FORMATION IN EGYPT AND CANARY ISLAND DESERTS

K. **Stahr** & R. **Jahn**, Stuttgart
A. **Huth** & J. **Gauer**, Berlin

Summary

The eolian transport of sediments is a dominant process in the Sahara desert and its surroundings. The eolian influence is exemplified in four characteristic soil profiles. In the Central Sahara profiles of the plateaus also produce dust. The loss of dust totals more than 300 kg/m^2 over some 20,000 years, at an average rate 0.1 mm/year. In the fringes of the desert, as well as in the isolated semi-desert of Lanzarote, the dust balance is positive. In all cases, a local eolian component occurs. The influence of the dust on soil development is different, depending on the location in the landscape. On pure limestones, silicates are mainly imported. In contrast, at the desert fringes on marls, the soil receives the lime content from dust, which forms calcretes. Finally, in soils derived from basalt (Lanzarote), the basic autigenic material is diluted by the dust import. Each profile has developed since the middle or young Pleistocene.

Zusammenfassung

Transport von Windsedimenten ist ein bedeutender Prozess in der Sahara und ihren Randgebieten. An vier Beispielen jeweils charakteristischer Profile wird der Einfluß äolischer Umlagerung dargestellt. Auch Hochflächenprofile sind in der Zentralsahara durch Auswehung beeinflußt. Die Verluste an Staub betragen mehr als 300 kg/m^2 bzw. mehr als 0.1 mm/a. In den Randbereichen der Wüste wie in der isolierten Halbwüste Lanzarotes überwiegt der Eintrag äolischer Sedimente. In jedem Fall tritt dabei auch ein Einfluß einer lokalen Komponente auf. Je nach der Ausgangssituation der Landschaft hat der Staub einen unterschiedlichen Einfluß auf die Bodenentwicklung. Auf reinen Kalksteinen ist die äolische Komponente hauptsächlich der Träger der Silikatminerale. Im Gegensatz dazu wird in der Randwüste und auf Mergeln, Kalk als Ausgangsmaterial der Kalkkrusten eingetragen. Auf die Böden aus Basalt in Lanzarote wirkt sich schließlich der Eintrag von Staub als Verdünnung aus. Alle Profile sind seit dem Mittel- und Jungpleistozän entstanden.

ISSN 0722-0723
ISBN 3-923381-17-4
©1989 by CATENA VERLAG,
D–3302 Cremlingen-Destedt, W. Germany
3-923381-17-4/89/5011851/US$ 2.00 + 0.25

1 Introduction

Many authors have contributed to the understanding of dust transport from the Sahara to the surrounding lands (YAALON & GANOR 1973, BENNETT 1980, GOUDIE 1983). It is well known that every year millions of tons of fine material leave the Sahara towards all directions. YAALON (1987) has recognized three different forms of transport. First is, the global long distance transport of mainly medium and fine silt fraction. The medium distance transport of 50–200 km of material less than 100 μm in diameter, in general, produces the desert fringes as well as the periglacial loess (ALAILY 1972, MAUS & STAHR 1977). Short distance transport (local loess) is reported in periglacial environments (ALAILY & PAPENFUSS 1973, MAUS & STAHR 1977) and includes saltation transport of sand size particles.

Two major questions appear to be unsolved. The first is that the rate of dust accretion is not only a function of meteorological factors (wind, rain) but also of the availability of dust and the condition of specific surfaces, which can trap the dust received to differing extents. The second problem concerns the possibility of tracing the dust. The mineralogy and chemistry of dust varies to a certain extent. Typical minerals like quartz or certain heavy minerals are not diagnostic in every case. The grain size distribution is fairly typical but if the dust is only a minor admixture, the eolian silt cannot be separated from the autigenic part.

In this paper the authors show, using simple methods for four selected examples, how variable the influence of eolian processes on the soil material and soil genesis can be.

2 Methods

- pH and electrical conductivity with 1:2,5 extract after 12 h, using WTW-electrodes.

- Particle size analysis according to SCHLICHTING & BLUME (1966) after H_2O_2 and HCL treatment.

- HF-$HCLO_4$ dissolution according to JACKSON (1960) for total element analysis. SiO_2 was calculated from the 100% minus the other oxides. Elements were determined by AAS and ICP (both Perkin Elmer) spectrometric methods.

- Conductometric analysis of carbonates after digestion with H_3PO_4 at 80°C and of carbon at 1000°C using Wösthoff apparatus.

- Amount of H_2O, OH^- and H_3O^+ by differential weight loss 105–550°C.

- Estimation of quartz on selected silt-samples, counting endothermic low$\Rightarrow$high -quartz inversion peak (573°C) with DTA.

- Mineral counting with petrography microscope on separated grain size fractions mainly using polarized light microscopy.

- Clay mineral analysis (fraction <2μm) of sediment samples with H_2O_2 treatment and cation exchange (with strong acid cation exchanger) down to pH 4-5, using X-ray diffractometer.

- Balance methods comparing parent material and current content (SCHLICHTING & BLUME 1966, ALAILY 1972, STAHR 1979 and JAHN 1988).

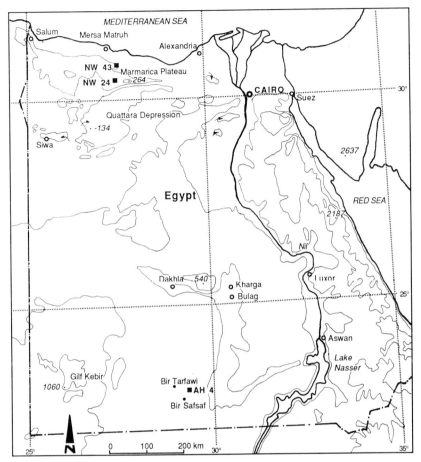

Fig. 1: *General map of Egypt. the location of profiles AH4 near Bir Safsaf, NW24 near Quaret el Shigara and NW43 on Haggag el Zarga are dotted.*

3 Soils of the Bir Safsaf-complex, southern Egypt

3.1 General

Southern Egypt is dominated by landscapes derived from different Nubian sandstone members (KLITZSCH & WYCISK 1987). The climate of the southern part of the western desert is hyperarid with less than 1 mm rainfall and about 24°C average annual temperature. Relative humidity is often as low as 10%. Rainfall observations between 1981 and 1986 registered no event with more than 0.1 mm.

North of the natural oasis of Bir Safsaf there is a dome of crystalline gneisses and granites with a diameter of about 50 km outcropping and inclining in all directions. This exposed, relatively homogeneous area offers the opportunity to study soil formation with a defined parent material, by chemistry, mineralogy and grain size. Fig.1 shows the general location of these sites and those of the Marmarica plateau mentioned in part 4.

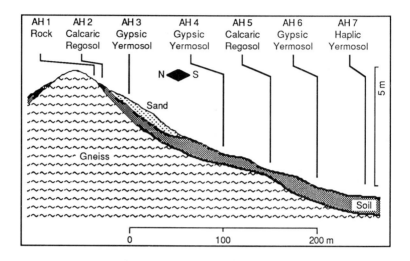

Fig. 2: *Typical soil catena within the gneiss area of Bir Safsaf basement complex.*

3.2 Soils

The soil association on the undulating plains, where frequent core stone rock outcrops occur, is characterized by the occurrence of Eutric and Calcaric Regosols, Orthic Solonchaks, Calcic and Haplic Yermosols. Lithosols and rock outcrops cover less than 10% of the landscape. There is no developed continuous drainage system. The depth of the soil cover over the solid rock reaches 80 to 100 cm but a lithic phase occurs frequently. Calcic and gypsic enrichments can be observed in many soils, but calcic horizons are restricted to flat depressions.

Fig.2 shows a selected catena with the profile discussed here. The whole area is affected by deflation. Old undisturbed surfaces with desert varnish and silicified pavement stones do not occur within the area in question, but can be found about 10 km to the north in old terraces also on the basement complex. The description of profile 4 is as follows:

Soil description of **Profile AH 4**

Location: 13 km NE of Bir Safsaf, Western desert, Egypt, about 300 m asl, without vegetation and landuse, exposed to the south, slope 1°, no groundwater contact.

Parent material: granitic gneiss.

Soil unit: Gypsic Yermosol, saline phase (FAO-UNESCO 1974).

Horizon description (March 1985)

0–2 cm E cover sediment, dry 7.5 YR 8/4, moist 7.5 YR 6/4, granular to structureless, loose consistence, pH 7.4, EC 0.37 mS, free of organic matter, very dry, abrupt boundary

2–8 cm Av* desert pan, dry 7.5 YR 6/4, moist 7.5 YR 5/4, medium size, platy, soft consistence, pH 7.5, EC 0.41 mS, no organic matter, very dry, abrupt boundary

8–20 cm Bw1 cambic horizon, color as above, medium size, columnar, slightly hard, pH 7.6, EC 1.20 mS, very dry, clear boundary

21–30 cm Bw2 cambic horizon continued, dry 7.5 YR 6/6, moist 7.5 YR 5/4, strong coarse prismatic, very hard, pH 7.8, EC 2.54 mS, very dry, clear boundary

30–47 cm BC subsoil, dry 10 YR 7/3, moist 10 YR 5/3, strong coherent, very hard, pH 7.7, EC 4.62 mS, very dry, few thin dead roots, lime spots, clear boundary

47–75 cm Cky subsoil, dry 10 YR 7/2, moist 10 YR 6/4, strong coherent with rock structure, pH 7.7, EC 36.4 mS, few thin dead roots, very dry, frequent soft lime and gypsum concretions, clear boundary

75– cm R rock, dry 10 YR 7/2, moist 10 YR 5/3, pH 7.8, EC 24.3 mS, rock structure, very hard, not penetrable by digging instruments

Cracks, abundant until 20 cm; decreasing in thickness and number with depth; very few having rock contact, filled with coarse and fine sand.

* v = vesicular pores

The profile shows a vesicular desert pan and well developed cracks as typical features of desert soils. Only a desert pavement is missing. This can be taken as a sign of relatively young development, since stones are available for the formation of a desert pavement and intensive salt weathering is not present, which could destroy the stones (DAN et al. 1982, ALAILY 1985). One of the major processes of soil formation, which can be recorded by the analytical data (compare profile description to fig.3), is physical weathering of the rock, with a maximum intensity at the topsoil. Chemical weathering results in a slight alkalization of the profile, carbonatization, gypsum formation and clay formation. The clay fraction consists of smectite > kaolinite ≫ illite. There is a relative decrease of smectite up the profile in favour of kaolinite. The enrichment of carbonates and gypsum cannot be explained by the weathering of the profile itself but requires one to assume some lateral additions from the close surrounding profiles 1–3. Profile data confirm the assumption.

The time of soil formation cannot be discussed in detail here. The results show that the profile was developed in an arid or strongly arid environment, as prevails today. The development may have been enhanced by moister conditions in the late Pleistocene and Holocene (WENDORF & SCHILD 1980, PACHUR et al. 1987). Features leading to the assumption of paleodevelopment under moist conditions were not found.

3.3 The evidence of eolian dust export

The area under study is a tpyical deflation area. The erosion cut into the basement after the removal of the basal clastics of the Nubian sandstone series. The hillocks of gneiss are partly unweathered and show only very thin weathering crusts and no varnish, but are obviously affected by the thermal cracking. The produced debris is not transported by water but partly taken up by wind or transported by the scarce event of sheet erosion, to the next depression. This is proven by the fact that runoff systems, such as wadis, are absent. Sediments in the depressions are less than 1 m thick thereby confirming the wind uptake of fine material.

The profile AH4 – Gypsic Yermosol is taken as an example of the processes prevailing over entire landscape. Grain size distribution does not show a major stratification boundary but rather a somewhat finer part in the Cky and then a gradual development toward finer sizes from BC to Av. In the latter horizon, fine sand is the dominant fraction. Mineral analysis shows that less than 10% of this fraction is rounded, and can be accepted as belonging to shifting sand. This may be the major filling material of the cracks. The other part is entirely derived from local gneiss. The major structural boundary is also gradual and occurs within the BC-horizon, where the rock structure disappears. Consequently,

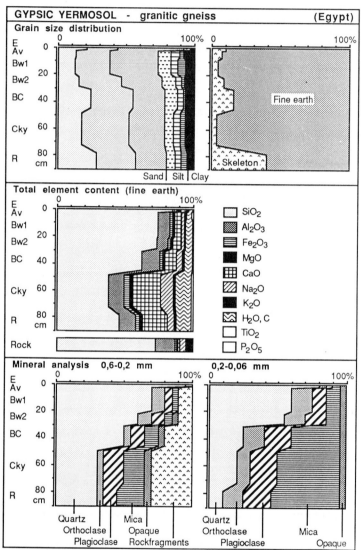

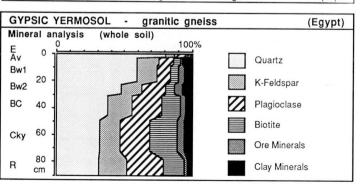

Fig. 3: top: Profile Gypsic Yermosol (AH4) near Bir Saf saf: depth function of skeleton, grain size distribution of silicatic fine earth, total element content of the whole fine earth and mineral analysis of medium and fine sand fraction.

Fig. 3: bottom: Depth function of the mineral distribution (% weight) on a whole soil basis for the Gypsic Yermosol (Profile AH4). Rock fragments and skeleton have been distributed among all fractions according to the rock contents.

soil formation in the lower part took place primarily by chemical and physical alteration without destroying the rock structure, whereas the upper part is completely reorganized into a new soil structure.

A striking feature of the total element and mineral depth functions (fig.3) is the sharp increase of silica and quartz, respectively, towards the top of the profile. It is shown in all profiles and for all grainsize fractions from coarse sand to silt. This can only be explained by a relative enrichment of quartz. Addition of quartz from the Nubian series can be estimated by the different morphology of the minerals. It is less than 10% in the top soil. Processes such as selective mineral weathering and translocation of elements cannot explain this phenomenon. Quartz enrichment can only be explained if micas, but mainly feldspars, disappear from the system. This loss could occur through direct uptake of silt particles by the wind, or water transport to the hollow 200 m away, and followed by wind uptake. Through this process of physical weathering and silt uptake, quartz must also be affected, but it is much more stable against weathering than the other rock-forming minerals.

The following calculation assumes a constant quartz amount in the profile and is therefore the minimal loss that could occur. Comparing the soil of the upper 4 horizons with the rock, gives 90 kg/m^2 more quartz than could be expected. Under the minimum conditions explained above, this surplus of quartz is equivalent to a loss of 330 kg/m^2 soil material. The original rock mass was 810 kg; 245 kg having been quartz, 485 kg feldspars, 65 kg mica and the rest opaques and heavy minerals. The quartz is, by definition, still the same.

The mica content is nearly as much as before. (Profile AH4 may have had a parent rock somewhat richer in micas than the average.) The major loss from the profile is a loss of feldspars. Only 95 kg are still present whereas 390 kg/m^2 disappeared or changed into clay minerals and carbonates (70 kg/m^2).

If this process continued only for the last period of 20,000 years, the average annual dust export would be more than 16.5 g/m^2/year or 20 cm in 20,000 years (100 μm/year). These assumptions do not explain how this process could affect a soil depth of what is now more than 30 and originally 50 cm. Possibly the soil material was reworked very slowly along the gentle slope, but this is not a convincing argument as the soil on the top shows the same feature, a bit less prominent.

4 Soils of the Marmarica Plateau, NW-Egypt

4.1 Landscape

Along a transect through the Marmarica Plateau (fig.1), the environmental conditions vary from a Mediterranean-influenced semi-desert in the north to a continental extreme desert at the Quattara depression in the south. Over the course of this 150 km transect, precipitation decreases from more than 150 mm to less than 20 mm, while mean annual temperature increases from 19 to 21°C.

The plateau is built up by limestone, dolomites and marls of the Miocene Marmarica formation, which was the terminal marine sedimentation. During the uplift of the Marmarica, a sequence of marine abrasion terraces were formed from Pliocene until recent times, stepping from 150 m down to the sea level

Distance (km)	3	10	15	20	30	35	40	50	90
Depth of calcretes (cm)	65	60	45	30	15	10	±	–	–
Thickness of calcretes (cm)		100		30		10	±	–	–
Powdery lime enrichment	+	+	+	+	+	+	(+)	(+)	–
CaCO$_3$-content of cover sand (%)	10	19	29	35	55	63	75	79	82

Tab. 1: *Change of soil properties of the Marmarica Plateau with distance from the sea.*

itself (SAID 1962).

From north to south the vegetation changes from diffuse to contracted, purely episodic vegetation (BORNKAMM & KEHL 1984).

Through pedological work, eight soilscapes have been found and described in a sequence from the Petrocalcic Xerosol-Luvic Xerosol-Fluvisol soilscape in the north, to the Lithosol-Regosol-Solonchak soilscape at the edge of the Quattara depression in the south. The sequence has a frame of Solonchaklandscapes along the shore and in the Quattara depression (GAUER & STAHR 1984, STAHR et al. 1985).

4.2 Soils

Two typical plateau soils have been selected in order to show the influence of eolian input on soil formation. The automorphic soils of the Marmarica change substantially from north to south. In the Mediterranean part lime redistribution, including extreme calcrete formation, clay illuviation and dune sand movement as well as wadi events, are the major soil forming processes.

The formation of calcic horizons occurs till 40 km inland; the visible redistribution of lime at about 60 km (tab.1). In contrast, gypsum and soluble salts increase their effect on soil formation. Clay illuviation continues with a smaller proportion of typical illuviated horizons at the end of the transect. Dune sand movement decreases southwards and continous wadi systems cut only 10–12 km inland. Takyric Solonchaks and Takyric Yermosols cover only minor parts of the surface but obtain their maximum distribution between 60–100 km from the sea. Lithosols are frequent on the top of the plateau and within the deeper parts of the southern decline. Salinity decreases sharply from the Mediterranean inland and then slowly increases toward the Quattara depression. Typical features of desert soils (DAN et al. 1982, ALAILY 1985) are best developed at the central part of the plateau between 20–50 mm annual rainfall. The two soils selected in this paper are on the northern higher plains (Pliocene marine terrace) and in the south of the central part of the plateau on a Miocene limestone surface.

Soil description of **Profile NW 24**

Location: Southern Marmarica Plateau, Western Desert, 107 km south from the Mediterranean coast along the abandoned telephone line from Wadi Garawla towards Siwa, 220 m asl., Plateau SW of Quaret el Shigara, vegetation cover 1% (*Hamada elegans, Zygophyllum album*), no land-use, no slope, no groundwater.

Parent material: Bioclastic hard dolomite, Marmarica formation, Miocene.

Soil unit: Luvic Yermosol, partly with lithic phase (FAO-UNESCO 1974).

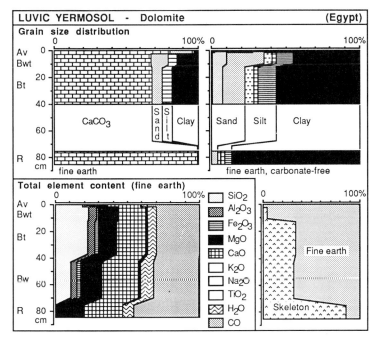

Fig. 4: *Profile Luvic Yermosol (NW24) near Quaret el Shigara: depth function of skeleton, grain size distribution of fine earth and silicatic fine earth as well as total element content of the whole fine earth.*

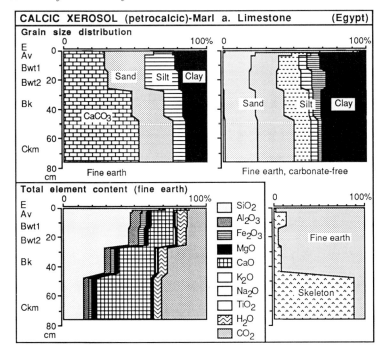

Fig. 5: *Profile Calcic Xerosol (petrocalcic phase) on the Hagag el Zarga plateau (NW43): depth function of skeleton and silicatic fine earth as well as total element content of the whole fine earth.*

Horizon description: (March 1982)

1–0 cm E cover sediment, about 80% coarse calcareous sand, few angular limestone fragments, snails

0–2 cm Av desert pan, dry 10 YR 8/3, moist 10 YR 7/4, coarse platy, vesicular pores, weak consistence, pH 7.5, EC 0.3 mS, 10% fine subangular limestones, free of organic matter, very dry, abrupt boundary

2–12 cm Bwt top soil, dry 10 YR 7/7, moist 7.5 YR 6/6, column to coarse subangular blocky, moderate consistence, pH 7.4, EC 0.2 mS, thin clay cutans in patches, very few angular limestones, very little organic matter, moist, clear boundary

12–40 cm Bt argillic horizon, moist 7.5 YR 6/6, medium subangular blocky, moderate consistence, pH 7.4, EC 0.25 mS, thin clay cutans in patches, 35% limestone fragments, very little organic matter, gradual boundary

40–70 cm Bw discontinuous fine earth horizon, moist 7.5 YR 5/6, coherent to subangular, pH 7.4, EC 0.25 mS, 30% coarse limestones in fine earth part, very few organic matter, stones with clay coatings above, moist, gradual broken boundary

>70 cm R hard bioclastic limestone, moist 7.5 YR 7/7, pores partly filled with reddish clay and hard white carbonate

Soil description of **Profile NW 43**

Location: Marmarica Plateau, Western Desert, Northern higher plains, Haggag el Zarga, 16.5 km along the abandoned telephone line from Wadi Garawla towards Siwa, 100 m asl, plain, 3.5 km north of a Pliocene cliff, semi-desert shrubs dominated by *Thymelea hirsuta* 25% cover, camel and goat pasture, no slope, no groundwater.

Parent material: Cover sediments above marls and limestones, Marmarica formation, Miocene.

Soil unit: Luvic Xerosol, petrocalcic phase (FAO-UNESCO 1974).

Horizon description: (March 1982)

1–0 cm E cover sand, moist 7.5 YR 7/7, single grained with few fragments of calcrete and snails

0–2 cm Av desert pan, dry 7.5 YR 7/5, moist 7.5 YR 6/5 coarse platy, typical vesicular pores, strong consistence, pH 8.0, EC 0.2 mS, few platy calcrete fragments, very little organic matter, dry, abrupt boundary

2–12 cm Bwt1 top soil, dry 7.5 YR 7/6, moist 7.5 YR 6/6, very coarse prismatic with subangular blocky fine structure, moderate consistence, pH 8.2, EC 0.3 mS, 15% calcrete fragments, very low organic matter (0.3%), dry, clear boundary

12–25 cm Bwt2 argillic horizon dry 7.5 YR 6.5/6, moist 7.5 YR 5.5/6, medium to fine subangular blocky, strong consistence, pH 8.2, EC 0.4 mS, few calcrete fragments, thin broken clay cutans, dry, clear boundary

25–45 cm Bkc calcic horizon, dry 7.5 YR 8/3, moist 7.5 YR 7.5/4, coherent to fine subangular blocky, strong consistence, pH 8.1, EC 0.40 mS, 10% coarse calcrete fragments, diffuse soft cementation with lime, dry, abrupt boundary

45– ≈ 75 cm, Ckm hard pan calcrete, dry, 7.5 YR 8/3, moist 7.5 YR 7/4, extremely hard upper part platy then massive, reddish-brown matrix with freqent snails incorporated, at the surface laminar crust, pH 8.1, EC 0.4 mS indurated cementation, dry.

Contact with underlying marl not exposed at the site but obvious through observations from nearby profiles.

Profile NW 24 (fig.4) is a typical plateau soil under (full) desert conditions. Those soils are fairly widespread in the area, but only on stable surfaces and NW 24 is there maximum development in thickness and expression of the morphological features. Especially the desert pan, with its vesicular pores and a regular net of sand-filled cracks, is obvious as a characteristic feature of this desert soil. From the field description, this is a limestone soil with clay illuviation, despite still being highly calcareous. The influence of salts is at a minimum and the 0.5% gypsum throughout the profile is very low.

Profile NW 43 (fig.5) is also a typical plateau soil but characterizes semi-desert conditions. The slope along the whole Haggag el Zarga plain is minimal. Flat, shallow depressions are filled with playa sediments and used as barley fields by the Bedouins. Very few are older karst hollows containing more weathered and reddened soil material than the recent soils. In such hollows until now the seepage is higher than the average. These are the only places where no calcretes develop. The overall existing calcretes have an average depth of 40 cm in the area, increasing towards the north and decreasing towards the south. The striking feature is that this correlates with the recent climate more than with the age of the marine terraces. Consequently, the thickness and depth of the calcrete is highest on the low-lying Pleistocene terraces.

Calcium carbonate enrichment begins right in the horizon above the calcrete; this is a general observation for the area. Two conclusions can be drawn from this feature. Firstly the calcrete requires several phases of solution and precipitation until its final cementation; secondly the calcic horizon above the calcrete may be a sign of increasing aridity when the regular moistening of the profile does not reach as deep down as in an earlier period.

4.3 Influence of silicates and lime addition on soil formation of the Marmarica Plateau

The Luvic Yermosol (NW 24) is an autigenic formation in the sense that there is no sign of colluvium or erosion. This is confirmed by the fact that the entire skeleton is defined from the bioclastic rock with some micritic parts. However, micromorphological examination shows that in all horizons above the rock, an equally distributed content of angular quartz grains of the silt fraction which are not found in the rock. The formation of fine earth from rock is not probable because the rock solution residue is much finer than that of the carbonate-free fine earth of the entire soil profile. Furthermore, an attempt to calculate the amount of dissolved limestone, in order to determine the silicate fine earth, results in 70 t/m^2 which is unlikely for such an aridic environment. Therefore we must assume an eolian input in order to form the entire fine earth.

The total enrichment of silicatic fine earth is 140 kg/m^2. Local loess found in shallow silty profiles contains 75% lime with about 2% sand, 80% silt and 18% clay. The grain size distribution of the fine earth, again, is not identical. The difference can be explained by the local addition of sandy (fine sand) material and by the slow decalcification which resulted in about 100–150 kg/m^2 of calcium carbonate loss. From this assumption the derived soil material is 60–70% dust transported from afar ($\approx$400 kg/m^2); the coarse material, about 25–30%, is locally transported fine sand (80% carbonates). The rest being derived from solution of the underlying rock and adding mainly clay minerals. The sand was added mainly to the top 12 cm and may have been reworked several times during soil formation. The entire period of soil formation is not known.

Given the knowledge gained from better-dated profiles further north and in the desert fringe areas of Israel we can assume a dust import of about 3 g/m^2/year, which would give a minimum age of the soil formation of 150,000 years. We accept that this is a very rough estimation but because there is nothing on which to base a valid absolute dating we cannot proceed further at the moment. We would not be surprised if the actual age is much older, because the surface was never very suitable for dust accretion (low vegetation density, absence of rough surface).

The Petrocalcic Xerosol (NW 43) has a different development but again the soil matrix has only minor relations to the underlying Miocene marls. On the other hand, the macro- and micromorphological observations, the grain size distribution as well as the chemical analysis, make it probable that the profile including the calcrete was derived from the same parent material. This parent material is typical of the northern part of Marmarica and may be called cover sediment. It is much more homogeneous than the underlying Miocene rocks. On the surface we frequently find terrestrial snails. The same snails are found through the profile right down into the calcrete. This favours the idea that the profile grew through a continuous addition of soil material as well as through a continuous translocation of calcium carbonate. Because the loess has a high carbonate content, it is also here the major source of lime (the marls only have 25–30% lime).

The total mass of this profile, including 30 cm of calcrete, is 1340 kg/m^2. this is mainly a mixture of local eolian sands (720 kg/m^2; 25% lime) and long-distance transported loess (520 kg/m^2; 75% lime, tab.1). This can explain the total amount of carbonate as well as the grain size of the fine earth. The entire skeleton was formed **in situ** (calcrete fragments). The average carbonate concentration of the mixture would be 55%. The lime which was added by the loess was mainly redistributed and now forms the calcrete horizon and the carbonate enrichment above. Thus, 325 kg of lime has been redistributed. Assuming a transport of 40 mg/l in the soil solution and 100 mm of rainfall, the yearly increment to the calcrete would be 4 g/m^2. This gives a minimum age of 80,000 years for the soil. Because the calcrete is partly broken and also forms the skeleton of the upper horizons, we have to assume that the lime was reprecipitated two or three times, giving a middle Pleistocene age for the whole formation. Given an age of 80,000 years, the rate of net loess input would be 6 g/m^2/year = 4μm/year.

A question that should be answered concerns the source of the dust and sand material. A major source of quartz sand could be from the Nubian series. This does not, however, fit the picture of sand distribution in Marmarica. The amount of sand sheets, dunes, and their quartz content, increases to the north. Therefore, the regional source must be the Mediterranean area. We assume that sand was transported from the Nile delta or from some wadis along the shore, and from there by wind. In contrast, the dust has too high a lime content to be produced in the north. The Quattara depression and the surrounding Eocene and Miocene limestones are very likely to be the source area. Dust storms in the Matruh area often have southerly wind directions. A comparison of these assumption with the weather records of Marsa Matruh shows that the main winds come from the NNW; the secondary direction in the frequency diagram is south. The average wind velocity there is 5.2 m/s. About 50% of the time the wind exceeds 5.4 m/s, being sufficient to take up silt and fine sand.

5 Soils of Lanzarote, Spain

5.1 Landscape

The island Lanzarote is formed from volcanic eruptive rocks, which have build up distinct land surfaces of different ages (fig.6). The age of the six basaltic se-

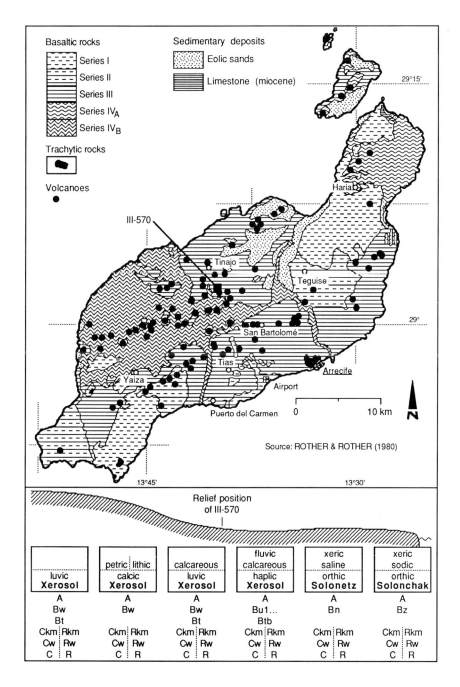

Fig. 6: *Geological map of Lanzarote and distribution of soils within an idealized catena on rocks of series III.*

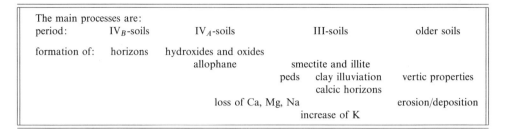

Tab. 2: *Dominant soil forming processes in Lanzarotes volcanic landscapes.*

ries covers the period from Miocene (series I), early to middle Pleistocene (series II_A, II_B), younger Pleistocene (series III), early Holocene (series IV_A), to the recent eruptions of 1730–36 and 1824 (series IV_B). The geochemistry of the rock types is basic to ultrabasic with small variations (FUSTER et al. 1968). The climate of the island (800 km², 29°N) is characterized by an average precipitation of ≈140 mm (<100–250) and mean annual temperatures of ≈20°C. The climate is classified as semiarid with warm temperatures.

The wind-regime is generally defined by the trade winds coming from the NE. In general, hot air masses from the Sahara, carrying dust, reach the Canary Islands for several days in the summer (FERNANDOPULLE 1976).

5.2 Soils

Depending on the age of the land surface, different stages of soil development can be found. On series IV_B, Eutric Regosols, derived from pyroclastic fall deposits, and bare rocks are found, whereas from pyroclastics of series IV_A Mollic Andosols have recently developed. Within older landscapes, Luvic Xerosols and Calcic Xerosols are widespread. For soils of the older series (older than III), polygenetic soil development is characteristic. Large areas of the older landscapes are affected by severe erosion and deposition of soil material in depressions (JAHN et al. 1983, JAHN 1988).

Different soil forming processes dominate the different stages of soil development found in Lanzarote. The main processes are shown in tab.2. (JAHN et al. 1985, JAHN & STAHR 1986, JAHN et al. 1987)

To compare the influence of different eolian dust, a Luvic Xerosol, derived from basalt of series III, was selected. The choosen profile, III-570, is part of a catena of Luvic Xerosols (plateau), petric/lithic Calcic Xerosols (upper slope), calcareous Luvic Xerosols (downhill and plain), and of fluvic, calcareous Haplic Xerosols (depression), representing the plain-position. Due to an even stone content this soil, from a plain but rough area, is considered as not affected by erosion or deposition of local soil material. However, the topsoil is influenced by recent agricultural practices such as ploughing and covering with a fine-gravelly lapilli layer.

Soil description of **Profile III-570**

Location: Island of Lanzarote, 500 m east of the Eremita de los Dolores (Tinguation), 270 m asl, exposed to east, slope <2°, vegetable gardening, no groundwater.

Parent material: tholeiitic basalt of the younger Pleistocene series III.

Soil unit: Luvic Xerosol, stony phase (FAO-UNESCO 1974).

Horizon description: (Oct. 1981)

10–0 cm Y fine gravelly surface layer of unweathered basaltic pyroclasts

0–11 cm Y/Ap silty-sandy loam, enriched with coarse material from the surface layer, 5 YR 4/3 (moist), coherent, poor in carbonates, OM 0.6%, $pH_{(KCl)}$ 7.7, EC 0.11 mS, clay consists of 70% illite, remains is smectite, abrupt boundary

11–30 cm Ah silty-clayey loam, very skeletal, 5 YR 4/3, subpolyedric, poor in carbonates, OM 1.8%, pH 7.3, EC 0.17 mS, clay consists of 45% illite, remains is kaolinite and smectite, clear boundary

30–50 cm Bw silty-clayey loam, very skeletal, 5 YR 4/3, subpolyedric, poor in carbonates, OM 1.1%, pH 7.0, EC 0.15 mS, clay consists of 40% smectite, remains is illite and kaolinite, gradual boundary

50–70 cm Bt1 argillic horizon, clay, very skeletal, 5 YR 3/3, subpolyedric to polyedric, clay cutans, poor in carbonates, OM 0.7%, pH 7.2, EC 0.39 mS, clay consists of 45% smectite, remains is illite and kaolinite, no boundary

70–95 cm Bt2 clay, very skeletal, 5 YR 3/3, subpolyedric to polyedric, clay cutans, carbonatic, OM 0.5%, pH 7.4, EC 0.52 mS, clay consists of 55% smectite, remains is illite and kaolinite, diffuse boundary

95–130 cm Bwk loamy clay, very skeletal, 5 YR 4/3, subpolyedric, rich in carbonates, OM 0.5%, pH 7.7, EC 0.41 mS, clay consists of 65% smectite, remains is illite and kaolinite

5.3 Addition of quartz and its influence on soil formation

From thin sections it is evident that a certain amount of quartz can be found in profile III-570, which cannot be explained by soil forming processes. This suggests external quartzy dust import. Furthermore, high silt and SiO_2 contents (fig.7) in the upper 50 cm of the profile also suggest a quartzy/silty component. Relative to the parent rock, an increase of 42 kg/m^2 of SiO_2 for the entire profile can be calculated (JAHN 1988).

On the other hand, slow soil development in the drier topsoil as well as clay illuviation can explain the high silt content of the topsoil. The content of SiO_2 can also be explained as relative increase through loss of about 150 kg/m^2 CaO, MgO and Na$_2$O. A loss of about 19 kg/m^2 SiO_2 was estimated using the Fe+Al/Si ratios from soil and rock.

The rough determination of quartz by DTA analysis leads to a quartz amount of about 40 kg/m^2 at depths 0 to 50 cm. By comparison, in these horizons 56 kg/m^2 of medium-silt, 34 kg/m^2 of fine-silt and 65 kg/m^2 of clay including 25 kg illite, were found. In the entire profile nearly 200 kg/m^2 of clay consisting of 90 kg smectite, 64 kg illite, 35 kg kaolinite and 10 kg oxides were calculated.

These amounts show that only a part of the clay fraction found in the profile can be explained as having been imported together with quartz. On the basis of a quartz/clay ratio of 1 for the dust, only 20% of the whole profile-clay can be explained as imported. In addition, only part of the illite can be considered as imported. Therefore, a new formation of illite from smectite must be possible (JAHN et al. 1987). However, the import of illite with eolian dust may be an important factor.

Assuming an age of series III of about 40,000 years, an annual increase of clay from dustfall of 1 g/m^2 is estimated, standing against a clay new-foramtion rate of 4 g/m^2/year (the age of the Quaternary basalts is defined by several beach terraces; see FUSTER et al. 1968). Focusing on the fine earth of the profile, which amounts to 520 kg/m^2 (all figures are calculated for material free of car-

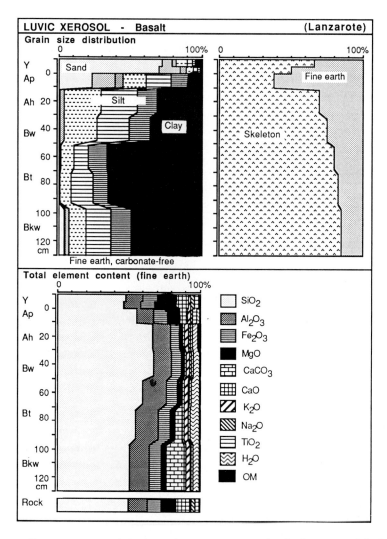

Fig. 7: *Profile Luvic Xerosol (III-570) at Lanzarote: depth function of skeleton and silicatic fine earth as well as total element content of the whole fine earth. Ca has been separated into silicatic CaO and carbonatic $CaCO_3$.*

bonates), the incorporated dust (total = 80 kg/m^2), about 15%, marks a remarkable, but not very great, role of eolian dust to the soil formation in Lanzarote.

6 Conclusions

An examination of four profiles in different deserts revealed that both dust production and aggregation differ in influence on soil formation at each site. The gneiss mounts of the Bir Safsaf crystalline complex, under extreme desert conditions, are a very efficient dust production area. Mainly feldspars and micas as well as an unknown part of the quartz, disappear. The soils on the Marmarica plateau, part desert on pure limestones, obtain their silicate component mainly from dust. The dissolution of limestone is not likely to produce as much silicate and the produced one has a different quality in comparison to the dust. The semi-desert soils of the Pliocene plateaus at the northern plains of NW Egypt mainly receive carbonates from the dust. In this case, virtually the entire soil material has been wind transported. Dust is one component, and dune sand, which comes from the north, being another source.

The Lanzarote soil, being separated from the Sahara desert by the Atlantic Ocean, receive only a minor portion of dust if dry eastern winds occur. Here, the dust adds silt quartz to the soil matrix together with part of the clay minerals, but no additional process is introduced by the silt. Therefore dust import is mainly a dilution.

This research has shown that the tracing of dust requires special methods for a specific site. In all cases, dust and local materials must be characterized by typical parameters. For the most part, the occurence and evidence of the dust, as well as the principal influence on soil formation, can be deduced.

A further question concerns the intensity and time of dust influence. Such an estimate would be highly speculative, but should be carried out in order to provide a general idea of the processes which produce soils. The soils described can be explained by formation in the late Pleistocene (Lanzarote and Bir Safsaf) to middle Pleistocene (Marmarica). Despite the uncertainties, formation since the tertiary period is unlikely.

Acknowledgements

This research is part of project "Lanzarote Soils", and special research project 69: "Geoscientific Problems in Arid Areas", both financed by the German Research Foundation (DFG). Without this support the fieldwork would not have been possible. We thank our colleagues of the General Petroleum Company of Egypt, Prof. Dr. H. Meshref from the University of Mansoura, Egypt, as well as Prof. Dr. E. Fernandez-Caldas and his group from La Laguna university, Spain, for their assistance, help and useful discussions.

References

ALAILY, F. (1972): Entstehung von Decklehmen auf Lias α-Schichtflächen in Südwestdeutschland und deren Veränderung durch Bodenbildung. Diss. Univ. Hohenheim, Stuttgart.

ALAILY, F. (1985): Nicht-hydrogene und hydrogene Böden — Merkmale in der Zentral Ostsahara. Mitteilungen Dtsch. Bodenkundl. Gesellsch., **43-II**, 705–710.

ALAILY, F. & PAPENFUSS, K.H. (1973): Entstehung von Decklehmen auf Lias Schichtflächen in Südwestdeutschland. Mitteilungen Dtsch. Bodenkundl. Gesellsch., **18**, 346–349.

BENNETT, J.G. (1980): Aeolian deposition and soil parent material in northern Nigeria. Geoderma, **24**, 241–256.

BORNKAMM, R. & KEHL, H. (1984): Pflanzengeographische Zonen in der Marmarica (Nordwest-Ägypten). Flora, **176**, 141–151.

DAN, J., YAALON, D.H., MOSHE, R. & NISSIM, S. (1982): Evolution of reg soils in southern Israel and Sinai. Geoderma, **28**, 173–202.

FAO-UNESCO (1974): Soil map of the world. Vol. I - Legend. Paris.

FERNANDOPULLE, D. (1976): Climatic characteristics of the Canary Islands. In: KUNKEL, G. (Ed.): Biogeography and Ecology in the Canary Islands. Monographiae Biologica, Bd **30**, Den Haag.

FUSTER, J.M., SANTIN FERNANDEZ, S. & SAGREDO, J. (1968): Geologia y Volcanologia de las Islas Canarias, Lanzarote. Inst. "Lucas Mallada", C.S.I.C., Madrid.

GAUER, J. & STAHR, K. (1984): Yermosol-Bodenlandschaften in NW-Ägypten. Mitteilungen Dtsch. Bodenkundl. Gesellsch., **39**, 19–24.

GOUDIE, A.S. (1984): Dust storms in space and time. Progress in Physical Geography, **7**, 502–530.

JACKSON, M.L. (Ed.) (1960): Soil Chemical Analysis, **2**. Prentice-Hall, Inc., Englewood Cliffs, N.Y.

JAHN, R. (1988): Vorkommen, Genese und Eigenschaften von Böden aus Vulkaniten im semiariden Klima Lanzarotes (Kanarische Inseln). Hohenheimer Arbeiten, Ulmer, Stuttgart.

JAHN, R., STAHR, K. & GUDMUNDSSON, Th. (1983): Bodenentwicklung aus tertiären bis holozänen Vulkaniten im semiariden Klima Lanzarotes (Kanarische Inseln). Z. Geomorph. N.F., Suppl.-Bd., **48**, 117–129.

JAHN, R., GUDMUNDSSON, Th. & STAHR, K. (1985): Carbonatisation as a soil forming process on soils from basic pyroclastic fall deposits on the Island of Lanzarote, Spain. CATENA SUPPLEMENT, **7**, 87–97.

JAHN, R. & STAHR, K. (1986): Changes in composition of elements within a chronosequence of soils by weathering of basic pyroclastics in the semiarid climate of Lanzarote. Transactions of the XIII. Congress of ISSS, Hamburg, Vol. III, 1152.

JAHN, R., ZAREI, M. & STAHR, K. (1987): Formation of clay minerals in soils developed from basic volcanic rock under semiarid climatic conditions in Lanzarote, Spain. CATENA **14**, 359–368.

KLITZSCH, E. & WYCISK, P. (1987): Geology of the sedimentary basins of northern Sudan and bordering areas. Berliner Geow. Abh., A **75.1**, 97–136.

MAUS, Hj. & STAHR, K. (1977): Auftreten und Verbreitung von Lößbeimengungen in periglazialen Schuttdecken des Schwarzwaldwestabfalls. CATENA, **3**, 369–386.

PACHUR, H.-J., RÖPER, H.-P., KRÖPELIN, S. & GOSCHIN, M. (1987): Late Quaternary hydrography of the eastern Sahara. Berl. Geow. Abh., A **75.2**, 331–384.

SAID, R. (1962): The Geology of Egypt. Elsevier, Amsterdam.

SCHLICHTING, E. & BLUME, H.-P. (1966): Bodenkundliches Praktikum. Hamburg, Berlin.

STAHR, K. (1979): Die Bedeutung periglazialer Deckschichten für Bodenbildung und Standortseigenschaften im Südschwarzwald. Freiburger Bodenkundl. Abhandlungen **9**.

STAHR, K., BORNKAMM, R., GAUER, J. & KEHL, H. (1985): Veränderung von Böden und Vegetation am Übergang von Halbwüste zur Vollwüste zwischen Mittelmeer und Quattara Depression in Ägypten. Geodynamik, Bd. **6**, 99–120.

WENDORF, F. & SCHILD, R. (1980): Prehistory of the Eastern Sahara. Academic Press, New York.

YAALON, D.H. (1987): Sahara dust and desert loess: effect on surrounding soils. Journal of African Earth Sciences **6** No. 4, 569–571.

YAALON, D.H. & GANOR, E. (1973): The influence of dust on soils during the Quaternary. Soil Science, **116**, 146–155.

Addresses of authors:
K. Stahr & R. Jahn
Universität Hohenheim
Institut für Bodenkunde und Standortslehre
D-7000 Stuttgart 70
West Germany
A. Huth & J. Gauer
Technische Universität Berlin
Department of Ecology — Soil Geography
Salzufer 12
D-1000 Berlin 10
West Germany

CLIMATIC AND NON-CLIMATIC CONTROLS OF ARIDITY: THE CASE OF THE NORTHERN NEGEV OF ISRAEL

A. **Yair** and S.M. **Berkowicz**, Jerusalem

Summary

The sensitivity of an ecosystem to climatic change may be greater for regions where loess soil-cover is predominant. As most climate modelling is usually conducted on the global scale, the effect of regional and more local climatic phenomena has been of secondary interest. The use of regional or local climatic data, such as temperature and rainfall, may still be insufficient to predict the outcome of climatic change for a particular region. Surface controls, such as the extent of soil and rock cover and, in addition, rainfall intensity will influence runoff and the degree to which water can be concentrated into a smaller area, making such an area potentially more fertile. The northern Negev of Israel, where two large regions of rocky and loess cover lie adjacent, is used to illustrate the above points. A research project presently being carried out in the northern Negev is described, and tentative models are offered as a means of correlating an arid/semiarid ecosystem to climatic and non-climatic environmental controls.

ISSN 0722-0723
ISBN 3-923381-17-4
©1989 by CATENA VERLAG,
D–3302 Cremlingen-Destedt, W. Germany
3-923381-17-4/89/5011851/US$ 2.00 + 0.25

1 Introduction

Climatologists use aridity indices to express the relationship between climatic variables and the environment. These indices tend to imply, however, that the acuteness of aridity in a given region depends upon prevailing atmospheric conditions, principally precipitation, and that aridity is inversely correlated to annual precipitation (WALLEN 1967).

This approach is certainly correct at the global scale, and even for annual crops where yields are greatly influenced by both annual precipitation and the temporal distribution of rainfall during the growing season, but it does not appear to fit the relationship between climate and environment at the desert fringe, if regional scales and longer time periods are considered.

A detailed palaeogeographical- ecological study of the evolution of the northern Negev, where annual rainfall now ranges from 70–250 mm, covering the last glacial period to the present, points to an uncertain relationship between annual precipitation and aridity (YAIR 1987, YAIR & ENZEL 1987). The degree of aridity, in terms of the evolution of the drainage network, soil-forming processes, composition of the vegetation, and human activity, indicates that environmental aridity for loess-covered areas increased during the relatively wet periods and decreased during the relatively drier periods. These processes can be explained by the important role that should be attributed to other factors controlling aridity, previously unconsidered, such as surface properties and the rainfall characteristics of the infrequent, individual storms.

In most of the arid zones of the world, life is largely dependent upon water supplied by direct rainfall, rather than from exogenous rivers or from meltwaters. In terms of a systems approach to water availability in arid regions, rainfall is the obvious input, with runoff, evaporation and percolation comprising the output. There is an interface however, which consists of surface properties, such as that of the soil-cover and the extent of bare-bedrock, and rainfall properties such as intensity, which will greatly influence the degree to which water will percolate or will be translated into runoff. It is the redistribution of water into certain places that can make an area more fertile. Therefore we believe that the biological productivity of plants and animals in the arid zone will not necessarily correlate with the actual amount of rainfall, but rather with the degree of increased water concentration, and that many soil properties, such as the depth of the gypsic and calcic horizons, and salinity profiles, will similarly be affected.

Accordingly, we contend that surface properties and rainfall characteristics encouraging water concentration can be considered as the main components of available water in the desert, rather than precipitation by itself. For this reason our comments are restricted to non-sandy areas as sand permits rapid and deeper water infiltration, thus inhibiting runoff.

2 Aridity indices and related relationships

It is useful, at this stage, to list some of the more commonly-used climatic classifications (fig.1). According to year of publication, they include DE MARTONNE (1926), KÖPPEN (1931) and subsequent revisions, THORNTHWAITE (1948) and the later refinement by THORNTHWAITE & MATHER (1955), EMBERGER (1955), BUDYKO (1956, 1974), BAGNOULS & GAUSSEN (1957), and BAILEY (1979).

The current and most widely accepted map showing the World's arid and semi-arid regions is that by UNESCO (1979), which uses the ratio of mean annual precipitation to mean annual potential evapotranspiration (P/ETP) as the basis for demarcation. Generally speaking, the indices have in common the following limitations:

1. The variables chosen tend to dwell upon temperature and precipitation averages

2. The criteria used to establish boundaries between climatic types are best-suited for the macro scale

3. Differentiation within the arid zone is not normally possible

Energy-based indices, notably that of BUDYKO (1974), still require a calculation which is based on the radiation needed to evaporate the annual precipitation.

Parallel to climatic classifications, ecological and pedological generalizations can be found in the literature. SHMIDA et al. (1986), for example, relates vegetation cover, productivity and number of species, to annual rainfall (fig.2). In so

DE MARTONNE (1926)

$A = \frac{P}{T+10}$

A = Aridity Index
P = mean annual precip (mm)
T = mean annual temp (°C)

EMBERGER (1955)

$Q = \frac{(100)(P)}{M^2 - m^2}$

Q = Pluviothermic Quotient
P = mean annual precip (mm)
M = mean max temp of hottest month (°C)
m = mean min temp of coldest month (°C)

KÖPPEN (1931)

$1 < \frac{P}{T} \leq 2$, and winter rain = steppe

$\frac{P}{T} \leq 1$, and winter rain = desert
$\frac{P}{T} > 2$, and winter rain = humid

P = mean annual precip (cm)
T = mean annual temp (°C)

BUDYKO (1956)

$D.I. = \frac{R}{L \cdot P}$

D.I. = Dryness Index
R   = Radiation balance
P   = Energy required to evaporate precipitation
L   = Latent heat of vaporization

$1 < D.I. < 2$ = Steppe
$2 < D.I. < 3$ = Semiarid
$D.I. > 3$ = Desert

THORNTHWAITE & MATHER (1955)

$I = 100/((p/PE)-1)$

I = Moisture Index
P = annual precip (mm)
PE = potential evapotranspiration (mm)

$I \geq -67$ = Arid
$-33 < I < -67$ = Semiarid
$I \leq -33$ = Subhumid/Humid

BAILEY (1979)

$M.I. = S = \sum_{i=1}^{12} Si$ and $Si = \frac{0.18P}{1.045^t}$

M.I. = Moisture index
Si   = monthly Moisture Index
P   = monthly precip (cm)
t   = mean monthly temp (°C)

$M.I. < 2.5$ = arid
$2.5 < M.I. \leq 4.7$ = semiarid
$4.7 < M.I. \leq 6.4$ = dry subhumid
$6.4 < M.I. \leq 8.7$ = moist subhumid

BAGNOULS & GAUSSEN (1957)

$P < 2T$ equals a dry month

P = mean monthly precip (mm)
T = mean monthly temp (°C)

1 to 8 dry months = Mediterranean
9 to 11 dry months = Hot Subdesert
12   dry months = Hot Desert

Fig. 1: *Some commonly-used climate classifications.*

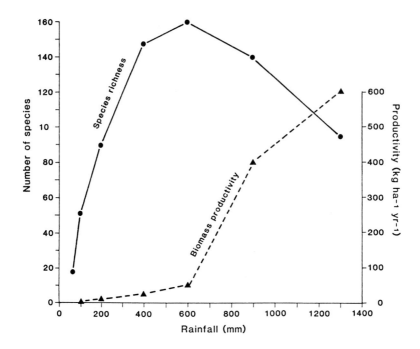
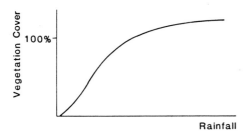

Fig. 2: *General relationships between vegetation variables and annual rainfall (modified from SHMIDA et al. 1986).*

far as general principles are concerned, as rainfall increases, so do the other variables. Pedological models which relate annual rainfall to the intensity of leaching, suggest that increasing annual rainfall leads to deeper leaching (fig.3).

The obvious reason why standard climate data are stressed in the abovementioned indices and models, is that they are meant to operate at a global scale. It is thus at the regional and local scales within the arid zone where their utility is particularly diminished. A salient example of this is provided in the northern Negev Desert of Israel.

3 A tale of two catchments

In the Beer Sheva Plains (fig.4) eolian loess deposits can be found lying directly on conglomerate or sedimentary rocks (YAIR & ENZEL 1987). This has led to the development of a continuous soil layer whose infiltration and water absorption capacity are high. This

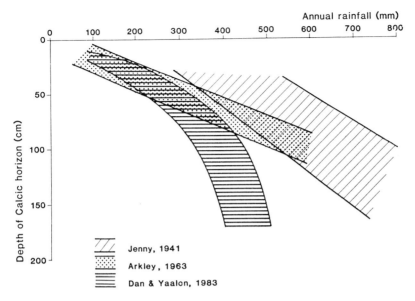

Fig. 3: *Relationship between annual rainfall and intensity of leaching processes (from YAIR 1987).*

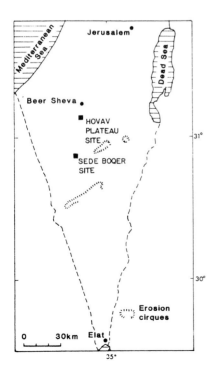

Fig. 4: *Location map.*

loess layer is affected by high evaporation leading to salt accumulation in the soil. Another area 30 km south however, adjacent to Sede Boqer in the Negev mountains, is characterized by a considerable proportion of bare rock with limited loess soil cover. This site experiences runoff which assists in concentrating rainfall into much smaller parts of the area (YAIR & LAVEE 1985), whithin which more water is available to permit deep infiltration and salt leaching.

In terms of hydrological, pedological, botanical and zoological considerations, it is the former area, in the Plains, that displays greater aridity. This is despite the fact that it receives an average of between 150–180 mm annual rainfall as opposed to 70–100 mm for the Negev mountains.

Tab.1 contains selected general climatic data for Sede Boqer and Beer Sheva.

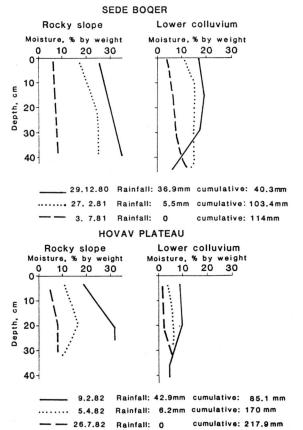

Fig. 5: *Moisture content at Sede Boqer and Hovav Plateau — rocky slope versus lower colluvium (from KARNIELI 1982 and KADMON 1984, respectively.*

The following figures compare and contrast data obtained from a variety of projects carried out in Sede Boqer and at a second experimental site in the Hovav Plateau, situated in the Plains just south of Beer Sheva. Soil moisture content data obtained on both the upper rocky and lower colluvium slopes are decidedly greater in Sede Boqer (fig.5). Soil property measurements (fig.6) of Na, Cl, and electrical conductivity at both sites illustrate the poorer leaching conditions for the Hovav Plateau (KARNIELI 1982, YAIR 1983, KADMON et al. 1989). It should be noted that even along the same slope, rocky areas show better conditions than their lower colluvial sections. A trench dug along a colluvium slope at Sede Boqer has in fact revealed that as one moves away from the rocky/colluvial interface, the calcic and gypsic horizons appear closer to the surface (fig.7).

If one compares floristic parameters of the two sites, especially of perennial vegetation, Sede Boqer once again reflects a relatively better environment for both rocky and colluvial slope sections (tab.2). Noteworthy is the relatively high percentage of Mediterranean vegetation found in Sede Boqer (YAIR & DANIN 1980, OLSVIG-WHITTACKER et al. 1983, KADMON et al. 1989).

	Distance from sea (km)	Elevation (m)	Mean min. (°C)	Mean max. (°C)	Mean temp. (°C)	Rain (mm)	Evaporation (mm)
Sede Boqer (34°47' E; 30°51' N)	70	480	12.1	24.7	18.7	93	2500
Beer Sheva (34°48' E; 31°48' N)	45	280	12.6	25.8	19.2	200	2250

Tab. 1: *Selected climatic data for Sede Boqer and for Beer Sheba.*

	Rocky section		Colluvial section	
	Sede Boqer	Hovav Plateau	Sede Boqer	Hovav Plateau
C	30[1]	10	10[1]	0
Sp	27[1]	11	no data	2
Sch	18[2]	7	14[2]	0
Smd	22[2]	18	7[2]	0
Ssa	33[2]	55	42[2]	100

C = Cover of perennial vegetation (%)
Sp = Number of perennial species
Sch = Number of semishrub species
Smd = Percentage of perennial species belonging to the Mediterranean, Mediterranean Irano-Turanian, and Mediterrranean Saharo-Arabian chorotypes
Ssa = Percentage of perennial Saharo-Arabian species
[1] after YAIR & DANIN (1980)
[2] after OLSVIG-WHITTAKER et al. (1983)
Hovav Plateau after KADMON et al. (1989)

Tab. 2: *Floristic comparisons — Sede Boqer and Hovav Plateau.*

Variable abundance per 100 m^2	Sede Boqer	Hovav Plateau
Number of isopod burrows	25	1
Number of porcupine digging points	30	0.2
Number of snails, *T. seetzenni*	260	6
Number of snails, *S. zonata*	20	20
Snail species richness	6	2

Tab. 3: *Zoological comparisons between Sede Boqer and Hovav Plateau (from YAIR & SHACHAK 1987).*

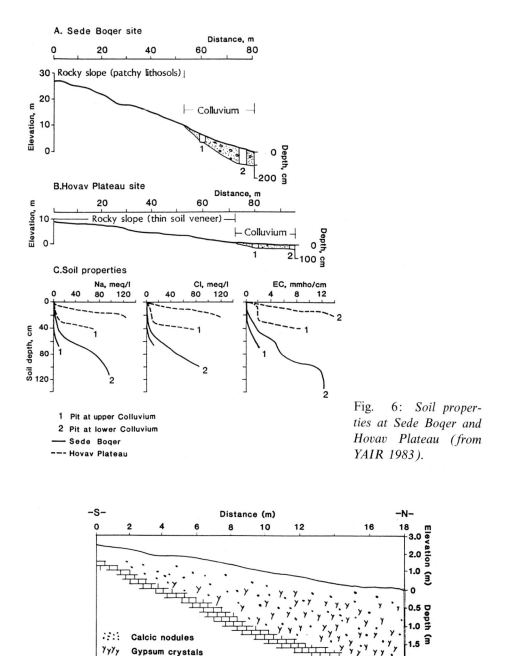

Fig. 6: *Soil properties at Sede Boqer and Hovav Plateau (from YAIR 1983).*

Fig. 7: *Occurrence of calcic nodules and gypsum crystals along a colluvium slope section at Sede Boqer (from WIEDER et al. 1985).*

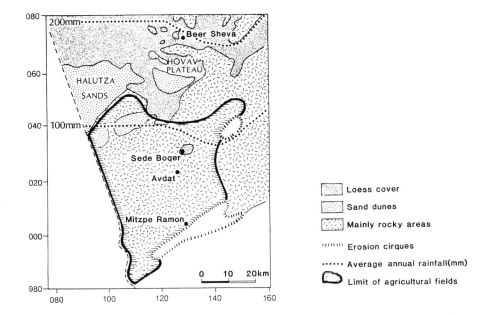

Fig. 8: *Areal distribution of ancient agricultural fields in the northern Negev (from YAIR 1983).*

Zoological indicators of environmental aridity show the same trend as for the pedological and botanical variables (tab.3). Porcupine, isopod and snail activity is considerably greater at Sede Boqer (YAIR & RUTIN 1981, YAIR & SHACHAK 1982). In addition, six snail species are present in the Sede Boqer site, of which 2 are species associated with a Mediterranean climate, versus only 2 species associated with a drier climate found at the Hovav Plateau (YAIR & SHACHAK 1987).

Furthermore, from the known limits of ancient agricultural fields in the northern Negev, it is evident that the mainly rocky areas were favoured as opposed to the higher rainfall, loess-covered regions (fig.8).

These inverse relationships suggest that standard climatic variables alone cannot adequately express the degree of aridity — indicating that there are other factors which have been overlooked. Overgrazing may be one of them, but it certainly cannot be used as an all-encompassing answer to account for the above observations.

We do not believe that the two research sites referred to above are the result of some anomalous local condition. To quote HEATHCOTE (1983):

"If the constraints of regional climate have any relevance for patterns of regional vegetation, then the existence of large areas of the world which have an arid vegetation type but only a semi-arid climate constraint are difficult to explain ... approximately 8.6% of the world's vegetation by area (12.5 million km^2) is in this way out of balance with its climate. In effect such areas appear to be poorer in quality than they ought to be ...".

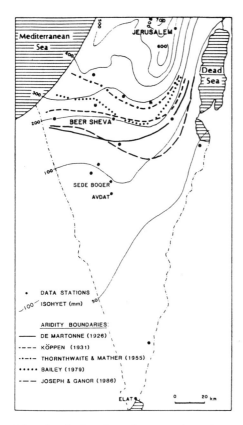

Fig. 9: *Aridity boundaries in Israel according to five climate classifications.*

4 Negev aridity index calculations

Given the environmental differences between Sede Boqer and the Hovav Plateau, we have applied 4 commonly-used climatic indices to the Negev and adjacent areas, in order to determine where the arid/semiarid boundaries would occur (fig.9). General climatic data were obtained from the Israel Bureau of Meteorology and other sources and cover periods ranging from 10 to 20 years. A fifth aridity boundary for Israel by JOSEPH & GANOR (1986) who use a modified Budyko-Lettau dryness ratio, has been superimposed in fig.9. The result, as expected, revealed the following trend — as one heads south, and with decreasing mean annual rainfall, the indices point to increasing aridity. Note that for each index in fig.9, the aridity boundary does not vary greatly and virtually all borders fall within 20 km of Beer Sheva.

5 On redefining aridity

Given the deficiencies of existing indices, and the environmental differences between the rocky Sede Boqer and loessic Hovav Plateau areas, an attempt is being made to refine an existing aridity index, or to develop a new index, with the purpose of obtaining a more suitable measure of the differing degrees of environmental aridity existing **within** arid/semiarid zones.

Our hypothesis behind the development of the index is the following:

For a non-sandy desert, aridity is controlled not only by temperature and rainfall but also by the bare bed-rock/soil-cover ratio and the nature of the rainfall regime. Under a given rainfall regime, water availability will be greater where this ratio is high. Accordingly, more leaching is expected as well as a decrease in soil salinity. In addition, the degree of biological productivity will be reflected by the presence of more mesic animal and plant species, with attendant higher population densities and biomass, and should be able to be expressed as a function of such physical parameters.

In effect we are proposing a new

paradigm of aridity by including those variables related to

a) the efficient transformation of rainfall into runoff water

b) variables enhancing water concentration which would aid biological productivity

c) climatic variables which have, to date, been little used

Three tentative predictive models are presented below using biological productivity variables, and physical characteristics, as indicators of the degree of aridity.

MODEL 1:
PB = f(MAR, RC, MAT, E, BBSC, EC, R)
MODEL 2:
AB = f(MAR, RC, MAT, E, BBSC, R)
MODEL 3:
D = f(MAR, RC, MAT, E, BBSC, R)

where PB = Plant biomass, AB = Animal biomass, and D = Depth to gypsic, calcic horizons, and
MAR = Mean annual rainfall
RC = Rain characteristics (intensity and duration)
MAT = Mean annual temperature
E = Evaporation
BBSC = Bare-bedrock/soil cover ratio
EC = Electrical conductivity at a given soil depth
R = Random element

Our procedure for testing the three models is as follows. Some 10 catchments have been selected in the northern part of the Negev, in which attention will focus on hillslopes of first and second-order streams, as a step towards isolating the variables to be incorporated in the models. The 300 mm isohyet forms the upper boundary of our study because of the degree of human interference and decreased bare-bedrock cover north of this line, with the 75 mm isohyet forming the approximate lower limit.

Within each catchment, a transect along the north-facing slope has been selected. Some transects consist of an upper rocky area which merges into a lower colluvial tongue, while others are predominantly loess-covered. The Sede Boqer and Hovav Plateau regions each have two study catchments in close proximity — one soil-covered and the other being rocky. Along the transects, measurements will be made of a number of variables known to reflect the degree of water availability. They are:

Pedological - depth to gypsic and calcic horizons
- electrical conductivity with depth

Botanical - % vegetation cover
- phytogeographic composition
- biomass per unit area

Zoological - isopod burrows per unit area
- porcupine digging points per unit area
- snails per unit area

The depth to the gypsic/calcic horizon, and electrical conductivity (salinity) measurements are useful indicators of water availability as they reflect the extent of leaching over a relatively long time-scale. Biomass is being determined by harvesting. Plant productivity measurements would have been preferable but data are difficult to acquire if several scattered sites need to be monitored. Due to their relatively limited mobility and the availability of existing data, isopod, snail and porcupine presence can serve as good indicators of the degree of aridity over relatively short time-scales.

Existing climate data from, or near, each study site are being analyzed. The range of standard data includes:

- monthly and annual temperature
- relative humidity
- monthly and annual rainfall
- potential evapotranspiration (calculated)

As runoff should be influenced by the rainfall intensity, rainfall data are being examined in order to statistically characterize rainfall for the study sites, and would include;

- rainfall intensity and duration of rainstorms as a whole
- rainfall intensity and duration of short showers
- frequency of rainfall intensities exceeding various thresholds
- number of showers per rainstorm
- time interval between consecutive showers
- rainfall energy

6 Concluding remarks

One may rightfully question whether refining an existing index, or the development of yet another classification, will be of practical benefit. We believe that a better understanding of the factors that determine aridity would provide a basis for defining the degree of sensitivity of a given arid/semiarid area to climatic fluctuations and, on the whole, a superior means of determining the agricultural potential of the arid zones. Furthermore, if it can be shown that rock-cover/soil ratios can serve as a simple guide to the degree of runoff, then such knowledge should permit improved guidelines to be developed for the management of rocky arid lands and, more specifically, lead to the delineation and mapping of areas best suited for agricultural development and other land uses such as grazing or recreational.

A salient example of this point is Sede Boqer (mean annual precipitation, 93 mm), where trees have been planted in small excavated minicatchments where the upper rocky and lower colluvium slope sections meet (YAIR & SHACHAK 1987). The runoff collected in the minicatchments has enabled these trees to now enter their fifth successful year of growth, without the benefit of irrigation, even after enduring two drought years.

It is clear that a refined aridity index, developed along the lines referred to earlier, could be of potentially great importance to Developing Countries situated in arid areas.

Acknowledgements

This resarch is being supported, in part, by the Arid Ecosystems Research Centre of the Hebrew University of Jerusalem and the Institute for Marine and Terrestrial Ecology (Toronto). SMB thanks the Lady Davis Fellowship Trust and the Moris M. Pulver Memorial Fund for assistance during the initial stages of the research. We thank Ms. T. Sofer, Ms. A. Bloch and Ms. M. Kidron for preparing the diagrams.

References

ARKLEY, R.J. (1963): Calculation of carbonate and water movement in soil from climatic data. Soil Science, **96**, 239–248.

BAGNOULS, F. & GAUSSEN, H. (1957): Les climats biologiques et leur classification. Annales de Géographie, **67 (355)**, 193–220.

BAILEY, H.P. (1979): Semi-arid climates: their definition and distribution. In: A.E. Hall, G.H. Cannell & H.W. Lawton (Eds.), Agriculture in Semi-Arid Environments. Springer-Verlag, Berlin, 73–96.

BUDYKO, M.I. (1956): The Heat Balance of the Earth's Surface. English translation N.A. Stepanova, 1958, U.S. Dept. of Commerce, Office of Technical Services, Washington.

BUDYKO, M.I. (1974): Climate and Life. Academic Press, New York.

DAN & YAALON (1982): Automorphine saline soils in Israel. CATENA SUPPLEMENT 1, 103–115.

EMBERGER, L. (1955): Une classification biogeographique des climats. Rec. Tr. Fac. Sci. L'Univ. Montpellier, Ser. Bot., 7, 3–43.

HEATHCOTE, R.L. (1983): The Arid Lands: Their Use and Abuse. Longman, New York.

JENNY, H. (1941): Factors of Soil Formation. McGraw-Hill, New York.

JOSEPH, J.H. & GANOR, E. (1986): Variability of climatic boundaries in Israel — use of a modified Budyko-Lettau aridity index. Journal of Climatology, 6, 69–82.

KADMON, R. (1984): Spatial variations in vegetation and environmental factors along arid slopes in the Hovav Plateau area. M.Sc. Thesis, The Hebrew University of Jerusalem. (in Hebrew with English summary).

KADMON, R., YAIR, A. & DANIN, A. (1989): Relationships between soil properties, soil moisture and vegetation along loess-covered hillslopes, northern Negev, Israel. CATENA SUPPLEMENT 14, 43–57.

KARNIELI, A. (1982): Spatial variation of soil moisture over arid hillslopes. M.Sc. Thesis, The Hebrew University of Jerusalem. (in Hebrew with English summary).

KÖPPEN, W. (1931): Die Klimate der Erde. Berlin.

DE MARTONNE, E. (1926): Une nouvelle fonction climatologique: l'indice d'aridité. La Météorologie, 2, 449–458.

OLSWIG-WHITTAKER, L., SHACHAK, M. & YAIR, A. (1983): Vegetation patterns related to environmental factors in a Negev desert watershed. Vegetatio, 54, 153–165.

SHMIDA, A., EVENARI, M. & NOY-MEIR, I. (1986): Hot desert ecosystems: an integrated view. In: M. Evenari, I. Noy-Meir & D.W. Goodall (Eds.), Hot Desert and Arid Shrublands (A). Elsevier, Amsterdam, 379–388.

THORNTHWAITE, C.W. (1948): An approach towards a rational classification of climate. Geographical Review, 30, 55–94.

THORNTHWAITE, C.W. & MATHER, J.R. (1955): The water balance. Publications in Climatology (Drexel Inst. of Tech.), 8, 1–104.

UNESCO (1979): Map of the world distribution of arid regions. MAB Technical Note No. 7, UNESCO, Paris.

WALLEN, C.C. (1967): Aridity definitions and their applicability. Geografiska Annaler (A), 49, 367–384.

WEIDER, A., YAIR, A. & ARZI, A. (1985): Catenary relationships on arid hillslopes. CATENA SUPPLEMENT 6, 41–57.

YAIR, A. (1983): Hillslope hydrology, water harvesting and areal distribution of some ancient agricultural systems, northern Negev, Israel. Journal of arid Environments, 6, 283–301.

YAIR, A. (1987): Environmental effects of loess penetration into the northern Negev desert. Journal of Arid Environments, 13, 9–24.

YAIR, A. & DANIN, A. (1980): Spatial variations in vegetation as related to the soil moisture regime over an arid limestone hillside, northern Negev, Israel. Oecologia (Berlin), 47, 83–88.

YAIR, A. & ENZEL, Y. (1987): The relationship between annual rainfall and sediment yield in arid and semi-arid areas. The case of the northern Negev. CATENS SUPPLEMENT, 10, 121–135.

YAIR, A. & LAVEE, H. (1985): Runoff generation in arid and semi-arid zones. In: M.G. Anderson & T.P. Burt (Eds.), Hydrological Forecasting. Wiley and Sons, New York, 183–220.

YAIR, A. & RUTIN, J. (1981): Aspects of the regional variation in the amount of available sediment produced by isopods and porcupines, northern Negev, Israel. Earth Surface Processes and Landforms, 6, 221–234.

YAIR, A. & SHACHAK, M. (1982): A case study of energy, water and soil flow chains in an arid ecosystem. Oecologia (Berlin), 54, 389–397.

YAIR, A. & SHACHAK, M. (1987): Studies in water shed ecology of an arid area. In: L. Berkovsky & G. Wurtele (Eds.), Progress in Desert research. Rowman and Littlefield, Totowa, New Jersey, 145–193.

Address of authors:
Aaron Yair & Simon M. Berkowicz
Department of Physical Geography
Institute of Earth Sciences
Hebrew University of Jerusalem
Givat Ram Campus
Jerusalem, Israel 91904

DAN H. YAALON (ED.)

ARIDIC SOILS and GEOMORPHIC PROCESSES

SELECTED PAPERS of the INTERNATIONAL CONFERENCE
of the INTERNATIONAL SOCIETY of SOIL SCIENCE

Jerusalem, Israel, March 29 – April 4, 1981

CATENA SUPPLEMENT 1, 1982

Price: DM 95,–/US $ 55.–
Special rate for subscription until June 30, 1982: DM 66,50/US $ 38.–
(available from the publisher only)
ISSN 0722–0723 / ISBN 3–923381–00–X

This CATENA SUPPLEMENT comprises 12 selected papers presented at the International Conference on Aridic Soils – Properties, Genesis and Management – held at Kiryat Anavim near Jerusalem, March 29 – April 4, 1981. The conference was sponsored by the Israel Society of Soil Science within the framework of activities of the International Society of Soil Science. Abstracts of papers and posters, and a tour guidebook which provides a review of the arid landscapes in Israel and a detailed record of its soil characteristics and properties (DAN et al. 1981) were published. Some 49 invited and contributed papers and 23 posters covering a wide range of subjects were presented at the conference sessions, followed by seven days of field excursions.

The present collection of 12 papers ranges from introductory general reviews to a number of detailed, process oriented, regional and local studies, related to the distribution of aridic soils and duricrusts in landscapes of three continents. It is followed by three papers on modelling and laboratory studies of geomorphic processes significant in aridic landscapes. It is rounded up by a methodological study of landform–vegetation relationships and a regional study of desertification. Additional papers, related to soil genesis in aridic regions, are being published in a special issue of the journal GEODERMA.

D.H. Yaalon
Editor

G.G.C. CLARIDGE & I.B. CAMPBELL
 A COMPARISON BETWEEN HOT AND COLD DESERT SOILS AND SOIL PROCESSES

R.L. GUTHRIE
 DISTRIBUTION OF GREAT GROUPS OF ARIDISOLS IN THE UNITED STATES

M.A. SUMMERFIELD
 DISTRIBUTION, NATURE AND PROBABLE GENESIS OF SILCRETE IN ARID AND SEMI–ARID SOUTHERN AFRICA

W.D. BLÜMEL
 CALCRETES IN NAMIBIA AND SE–SPAIN RELATIONS TO SUBSTRATUM, SOIL FORMATION AND GEOMORPHIC FACTORS

E.G. HALLSWORTH, J.A. BEATTIE & W.E. DARLEY
 FORMATION OF SOILS IN AN ARIDIC ENVIRONMENT WESTERN NEW SOUTH WALES, AUSTRALIA

J. DAN & D.H. YAALON
 AUTOMORPHIC SALINE SOILS IN ISRAEL

R. ZAIDENBERG, J. DAN & H. KOYUMDJISKY
 THE INFLUENCE OF PARENT MATERIAL, RELIEF AND EXPOSURE ON SOIL FORMATION IN THE ARID REGION OF EASTERN SAMARIA

J. SAVAT
 COMMON AND UNCOMMON SELECTIVITY IN THE PROCESS OF FLUID TRANSPORTATION:
 FIELD OBSERVATIONS AND LABORATORY EXPERIMENTS ON BARE SURFACES

M. LOGIE
 INFLUENCE OF ROUGHNESS ELEMENTS AND SOIL MOISTURE ON THE RESISTANCE OF SAND TO WIND EROSION

M.I. WHITNEY & J.F. SPLETTSTOESSER
 VENTIFACTS AND THEIR FORMATION: DARWIN MOUNTAINS, ANTARCTICA

M.B. SATTERWHITE & J. EHLEN
 LANDFORM–VEGETATION RELATIONSHIPS IN THE NORTHERN CHIHUAHUAN DESERT

H.K. BARTH
 ACCELERATED EROSION OF FOSSIL DUNES IN THE GOURMA REGION (MALI) AS A MANIFESTATION OF DESERTIFICATION

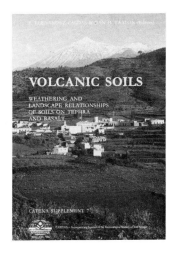

E. Fernandez Caldas & Dan H. Yaalon (Editors):

VOLCANIC SOILS
Weathering and Landscape
Relationships of Soils on Tephra and Basalt

CATENA SUPPLEMENT 7, 1985

Price DM 128,— / US $ 75.—

ISSN 0722–0723 / ISBN 3-923381-06-9

PREFACE

This CATENA SUPPLEMENT contains selected papers presented at the International Meeting on Volcanic Soils held in Tenerife, July 1984. The meeting brought together over 80 scientists from 21 countries, with interest in the origin, nature and properties of soils on tephra and basaltic parent materials and their management. Some 51 invited and contributed papers and 8 posters were presented on a wide range of subjects related to volcanic soils, many of them dealing with weathering and landscape relationships. Classification was also discussed extensively during a six day excursion of the islands of La Palma, Gomera and Lanzarote, which enabled the participants to see the most representative volcanic soils of the Canary Archipelago under a considerable range of climatic regimes and parent material ages.

Because volcanic soils are not a common occurrence in regions where pedology developed and progressed during its early stages, recognition of their specific properties made an impact only in the late forties. The name **Ando** soils, now recognized as a special Great Group in all comprehensive soil classification systems, was coined in 1947 during reconnaissance soil surveys in Japan made by American soil scientists. Subsequently a Meeting on the Classification and Correlation of Soils from Volcanic Ash, sponsored by FAO and UNESCO, was held in Tokyo, Japan, in 1964, in preparation for the Soil Map of the World. This was followed by meetings of a Panel on Volcanic Ash Soils in Latin America, Turrialba, Costa Rica, in 1969 and a second meeting in Pasto, Colombia, in 1972. At the International Conference on Soils with Variable Charge, Palmerston, New Zealand, 1981, the subject of Andosols was discussed intensively. Most recently the definitions of Andepts, as presented in the 1975 U.S. Soil Taxonomy, prompted the establishment of an International Committee on the Classification of Andisols (ICOMAND), chaired by M. Leamy from C.S.I.R., New Zealand, which held a number of international classification workshops, the latest in Chile and Ecuador, in January 1984. The continuous efforts to improve and revise the new classification of these soils is also reflected in some of the papers in this volume.

While Andosols or Andisols formed on tephra (volcanic ash), essentially characterized by low bulk density (less than 0.9 g/cm^3) and a surface complex dominated by active Al, cover worldwide an area of about 100 million hectares (0.8% of the total land area), the vast basaltic plateaus and their associated soils cover worldwide an even greater area, frequently with complex age and landscape relationships. While these soils do not generally belong to the ando group, their pedogenetic pathways are also strongly influenced by the nature and physical properties of the basalt rock. The papers in this volume cannot cover the wide variety of properties of the soils in all these areas, some of which have been reviewed at previous meetings. In this volume there is a certain emphasis on some of the less frequently studied environments and on methods of study and characterization as a means to advance the recognition and classification of these soils.

The Tenerife meeting was sponsored by a number of national and international organizations, including the Autonomous Government of the Canary Islands, the Institute of Ibero American Cooperation in Madrid, the Directorate on Scientific Policy of the Ministry of Education and Science, Madrid, the International Soil Science Society, ORSTOM of France, and ICOMAND. Members and staff of the Department of Soil Science of the University of La Laguna had the actual task of organizing the meeting and the field trips. In editing the book we benefitted from the manuscript reviews by many of our colleagues all over the world, and the capable handling and sponsorship of the CATENA VERLAG. To all those who have extended their help we wish to express warm thanks.

La Laguna and Jerusalem,   E. Fernandez Caldas
Summer 1984   D.H. Yaalon
     Editors

CONTENTS

R. PARFITT & A.D. WILSON
ESTIMATION OF ALLOPHANE AND HALLOYSITE IN THREE SEQUENCES OF VOLCANIC SOILS, NEW ZEALAND

A. HERNANDEZ MORENO, V. CUBAS GARCIA, A. GONZALEZ BATISTA & E. FERNANDEZ CALDAS
STUDY OF AMMONIUM OXALATE REACTIVITY AT pH 6.3 (Ro) IN DIFFERENT TYPES OF SOILS WITH VARIABLE CHARGE. I

FERNANDEZ CALDAS, J. HERNANDEZ MORENO, L. TEJEDOR SALGUERO, A. GONZALEZ BATISTA & V. CUBAS GARCIA
BEHAVIOUR OF OXALATE REACTIVITY (Ro) IN DIFFERENT TYPES OF ANDISOLS. II

J. RADCLIFFE & G.P. GILLMAN
SURFACE CHARGE CHARACTERISTICS OF VOLCANIC ASH SOILS FROM THE SOUTHERN HIGHLANDS OF PAPUA NEW GUINEA

GONZALEZ BONMATI, M.P. VERA GOMEZ & J.E. GARCIA HERNANDEZ
KINETIC STUDY OF THE EXPERIMENTAL WEATHERING OF AUGITE AT DIFFERENT TEMPERATURES

R. RIEZEBOS
HIGH–CONCENTRATION LEVELS OF HEAVY MINERALS IN TWO VOLCANIC SOILS FROM COLOMBIA: A POSSIBLE PALEOENVIRONMENTAL INTERPRETATION

L.J. EVANS & W. CHESWORTH
THE WEATHERING OF BASALT IN AN ARCTIC ENVIRONMENT

R. JAHN, Th. GUDMUNDSSON & K. STAHR
CARBONATISATION AS A SOIL FORMING PROCESS ON SOILS FROM BASIC PYROCLASTIC FALL DEPOSITS ON THE ISLAND OF LANZAROTE, SPAIN

P. QUANTIN
CHARACTERISTICS OF THE VANUATU ANDOSOLS

P. QUANTIN, B. DABIN, A. BOULEAU, L. LULLI & D. BIDINI
CHARACTERISTICS AND GENESIS OF TWO ANDOSOLS IN CENTRAL ITALY

A. LIMBIRD
GENESIS OF SOILS AFFECTED BY DISCRETE VOLCANIC ASH INCLUSIONS, ALBERTA, CANADA

M.L. TEJEDOR SALGUERO, C. JIMENEZ MENDOZA, A. RODRIGUEZ RODRIGUEZ & E. FERNANDEZ CALDAS
POLYGENESIS ON DEEPLY WEATHERED PLIOCENE BASALT, GOMERA (CANARY ISLANDS): FROM FERRALLITIZATION TO SALINIZATION

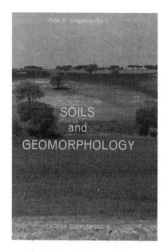

Peter D. Jungerius (Ed.):

Soils and Geomorphology

CATENA SUPPLEMENT 6 (1985)

Price DM 120,— / US $ 70.—

ISSN 0722–0723 / ISBN 3-923381-05-0

It was 12 years ago that CATENA's first issue was published with its ambitious subtitle "Interdisciplinary Journal of Geomorphology – Hydrology – Pedology". Out of the nearly one hundred papers that have been published in the regular issues since then, one-third have been concerned with subjects of a combined geomorphological and pedological nature. Last year it was decided to devote SUPPLEMENT 6 to the integration of these two disciplines. Apart from assembling a number of papers which are representative of the integrated approach, I have taken the opportunity to evaluate the character of the integration in an introductory paper. I have not attempted to cover the whole bibliography on the subject: an on-line consultation of the Georef files carried out on 29th October, 1984, produced 3627 titles under the combined keywords 'geomorphology' and 'soils'. Rather, I have made use of the ample material published in CATENA to emphasize certain points.

In spite of the fact that land forms as well as soils are largely formed by the same environmental factors, geomorphology and pedology have different roots and have developed along different lines. Papers which truly emanate the two lines of thinking are therefore relatively rare. This is regrettable because grafting the methodology of the one discipline onto research topics of the other often adds a new dimension to the framework in which the research is carried out. It is the aim of this SUPPLEMENT to stimulate the cross-fertilization of the two disciplines.

The papers are grouped into 5 categories: 1) the response of soil to erosion processes, 2) soils and slope development, 3) soils and land forms, 4) the age of soils and land forms, and 5) weathering (including karst).

P.D. Jungerius

P.D. JUNGERIUS
  SOILS AND GEOMORPHOLOGY

The response of soil to erosion processes
C.H. QUANSAH
  THE EFFECT OF SOIL TYPE, SLOPE, FLOWRATE AND THEIR INTERACTIONS ON DETACHMENT BY OVERLAND FLOW WITH AND WITHOUT RAIN
D.L. JOHNSON
  SOIL THICKNESS PROCESSES

Soils and slope development
M. WIEDER, A. YAIR & A. ARZI
  CATENARY SOIL RELATIONSHIPS ON ARID HILLSLOPES
D.C. MARRON
  COLLUVIUM IN BEDROCK HOLLOWS ON STEEP SLOPES, REDWOOD CREEK DRAINAGE BASIN, NORTHWESTERN CALIFORNIA

Soil and landforms
D.J. BRIGGS & E.K. SHISHIRA
  SOIL VARIABILITY IN GEOMORPHOLOGICALLY DEFINED SURVEY UNITS IN THE ALBUDEITE AREA OF MURCIA PROVINCE, SPAIN

C.B. CRAMPTON
  COMPACTED SOIL HORIZONS IN WESTERN CANADA

The age of soils and landforms
D.C. VAN DIJK
  SOIL GEOMORPHIC HISTORY OF THE TARA CLAY PLAINS S.E. QUEENSLAND
H. WIECHMANN & H. ZEPP
  ZUR MORPHOGENETISCHEN BEDEUTUNG DER GRAULEHME IN DER NORDEIFEL
M.J. GUCCIONE
  QUANTITATIVE ESTIMATES OF CLAY–MINERAL ALTERATION IN A SOIL CHRONOSEQUENCE IN MISSOURI, U.S.A.

Weathering (including Karst)
A.W. MANN & C.D. OLLIER
  CHEMICAL DIFFUSION AND FERRICRETE FORMATION
M. GAIFFE & S. BRUCKERT
  ANALYSE DES TRANSPORTS DE MATIERES ET DES PROCESSUS PEDOGENETIQUES IMPLIQUES DANS LES CHAINES DE SOLS DU KARST JURASSIEN

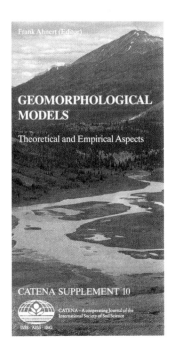

Frank Ahnert (Editor):

GEOMORPHOLOGICAL MODELS

Theoretical and Empirical Aspects

CATENA SUPPLEMENT 10, 1987

Price DM 149, — / US $88. —

ISSN 0722-0723 / ISBN 3-923381-10-7

CONTENTS

Preface

I. SLOPE PROCESSES AND SLOPE FORM

KIRKBY, M.J.
Modelling some influences of soil erosion, landslides and valley gradient on drainage density and hollow development.

TORRI, D.
A theoretical study of SOIL DETACHABILITY.

AI, N. & MIAO, T.
A model of progressive slope failure under the effect of the neotectonic stress field.

AHNERT, F.
Process-response models of denudation at different spatial scales.

SCHMIDT, K.-H.
Factors influencing structural landform dynamics on the Colorado Plateau – about the necessity of calibrating theoretical models by empirical data.

DE PLOEY, J. & POESEN, J.
Some reflections on modelling hillslope processes.

II. CHANNELS AND CHANNEL PROCESSES

SCHICK, A.P., HASSAN, M.A. & LEKACH, J.
A vertical exchange model for coarse bedload movement-numerical considerations.

ERGENZINGER, P.
Chaos and order – the channel geometry of gravel bed braided rivers.

BAND, L.E.
Lateral Migration of stream channels.

WIECZOREK, U.
A mathematical model for the geometry of meander bends.

III. SEDIMENT YIELD

YAIR, A. & ENZEL, Y.
The relationship between annual rainfall and sediment yield in arid and semi-arid areas. The case of the northern Negev.

ICHIM, I. & RADOANE, M.
A multivariate statistical analysis of sediment yield and prediction in Romania.

RAWAT, J.S.
Modelling of water and sediment budget: concepts and strategies.

MILLER, TH.K.
Some preliminary latent variable models of stream sediment and discharge characteristics.

IV. GENERAL CONSIDERATIONS

HARDISTY, J.
The transport response function and relaxation time in geomorphic modelling.

HAIGH, M.J.
The holon – hierarchy theory and landscape research.

TROFIMOV, A.M.
On the problem of geomorphological prediction.

SCHEIDEGGER, A.E.
The fundamental principles of landscape evolution.

Anton C. Imeson & Maria Sala:

GEOMORPHIC PROCESSES

In Environments With Strong
Seasonal Contrasts
Vol. I: HILLSLOPE PROCESSES

CATENA SUPPLEMENT 12, 1988

Price: DM 149, — / US $88. —

ISSN 0722-0723 / ISBN 3-923881-12-3

CONTENTS

Preface

A. Ávila & F. Rodá
Export of Dissolved Elements in an Evergreen-Oak Forested Watershed in the Montseny Mountains (NE Spain)

M. Sala
Slope Runoff and Sediment Production in Two Mediterranean Mountain Environments

J. Sevink
Soil Organic Horizons of Mediterranean Forest Soils in NE-Catalonia (Spain): Their Characteristics and Significance for Hillslope Runoff, and Effects of Management and Fire

A.G. Brown
Soil Development and Geomorphic Processes in a Chaparral Watershed: Rattlesnake Canyon, S. California, USA

T.P. Burt
Seasonality of Subsurface Flow and Nitrate Leaching

K. Rögner
Measurements of Cavernous Weathering at Machtesh Hagadol (Negev, Israel) A Semiquantitative Study

M. Mietton
Mesures Continués des Températures dans le Socle Granitique en Region Soudanienne (Fèvrier 1982–Juin 1983, Ouagadougou, Burkina Faso)

N. La Roca Cervigón & A. Calvo-Cases
Slope Evolution by Mass Movements and Surface Wash (Valls d'Alcoi, Alicante, Spain)

A. Calvo-Cases & N. La Roca Cervigón
Slope Form and Soil Erosion on Calcareous Slopes (Serra Grossa, Valencia)

J. Poesen & D. Torri
The Effect of Cup Size on Splash Detachment and Transport Measurements
Part I: Field Measurements

D. Torri & J. Poesen
The Effect of Cup Size on Splash Detachment and Transport Measurements
Part II: Theoretical Approach

A.C. Imeson & J.M. Verstraten
Rills on Badland Slopes: A Physico-Chemically Controlled Phenomenon

L.A. Lewis
Measurement and Assessment of Soil Loss in Rwanda

C. Zanchi
Soil Loss and Seasonal Variation of Erodibility in Two Soils with Different Texture in the Mugello Valley in Central Italy

L. Góczán & A. Kertész
Some Results of Soil Erosion Monitoring at a Large-Scale Farming Experimental Station in Hungary

H. Lavee
Geomorphic Factors in Locating Sites for Toxic Waste Disposal

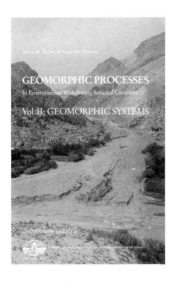

Adrian M. Harvey & Maria Sala:

GEOMORPHIC PROCESSES

In Environments With Strong
Seasonal Contrasts
Vol. II: GEOMORPHIC SYSTEMS

CATENA SUPPLEMENT 13, 1988

Price: DM 126, — / US $74. —

ISSN 0722-0723 / ISBN 3-923381-13-1

CONTENTS

Preface

M.A. Romero-Díaz, F. López-Bermúdez, J.B. Thornes, C.F. Francis & G.C. Fisher
Variability of Overland Flow Erosion Rates in a Semi-arid Mediterranean Environment under Matorral Cover, Murcia, Spain

R.B. Bryan, I.A. Campbell & R.A. Sutherland
Fluvial Geomorphic Processes in Semi-arid Ephemeral Catchments in Kenya and Canada

N. Clotet-Perarnau, F. Gallart & C. Balasch
Medium-term Erosion Rates in a Small Scarcely Vegetated Catchment in the Pyrenees

M. Gutiérrez, G. Benito & J. Rodríguez
Piping in Badland Areas of the Middle Ebro Basin, Spain

H. Suwa & S. Okuda
Seasonal Variation of Erosional Processes in the Kamikamihori Valley of Mt. Yakedake, Northern Japan Alps

F. Gallart & N. Clotet-Perarnau
Some Aspects of the Geomorphic Processes Triggered by an Extreme Rainfall Event: The November 1982 Flood in the Eastern Pyrenees

P. Ergenzinger
Regional Erosion: Rates and Scale Problems in the Buonamico Basin, Calabria

M. Sorriso-Valvo
Landslide-related Fans in Calabria

A.M. Harvey
Controls of Alluvial Fan Development: The Alluvial Fans of the Sierra de Carrascoy, Murcia, Spain

C. Sancho, M. Gutiérrez, J.L. Peña & F. Burillo
A Quantitative Approach to Scarp Retreat Starting from Triangular Slope Facets, Central Ebro Basin, Spain

A.J. Conacher
The Geomorphic Significance of Process Measurements in an Ancient Landscape

F. Ahnert, H. Rohdenburg & A. Semmel:

BEITRÄGE ZUR GEOMORPHOLOGIE DER TROPEN (OSTAFRIKA, BRASILIEN, ZENTRAL- UND WESTAFRIKA) CONTRIBUTIONS TO TROPICAL GEOMORPHOLOGY

CATENA SUPPLEMENT 2, 1982
Price: DM 120,–/US $ 69.–
ISSN 0722–0723 / ISBN 3-923381-01-8

F. AHNERT
UNTERSUCHUNGEN ÜBER DAS MORPHOKLIMA
UND DIE MORPHOLOGIE DES
INSELBERGGEBIETES VON MACHAKOS, KENIA

(INVESTIGATIONS ON THE MORPHOCLIMATE
AND ON THE MORPHOLOGY OF THE
INSELBERG REGION OF MACHAKOS, KENIA)

S. 1–72

H. ROHDENBURG
GEOMORPHOLOGISCH–BODENSTRATIGRAPHISCHER
VERGLEICH ZWISCHEN DEM
NORDOSTBRASILIANISCHEN TROCKENGEBIET
UND IMMERFEUCHT–TROPISCHEN GEBIETEN
SÜDBRASILIENS

MIT AUSFÜHRUNGEN ZUM PROBLEMKREIS DER
PEDIPLAIN–PEDIMENT–TERRASSENTREPPEN

S. 73–122

A. SEMMEL
CATENEN DER FEUCHTEN TROPEN
UND FRAGEN IHRER GEOMORPHOLOGISCHEN
DEUTUNG

S. 123–140

H.-R. BORK u. W. RICKEN

BODENEROSION, HOLOZAENE UND PLEISTOZAENE BODENENTWICKLUNG

SOIL EROSION, HOLOCENE AND PLEISTOCENE SOIL DEVELOPMENT

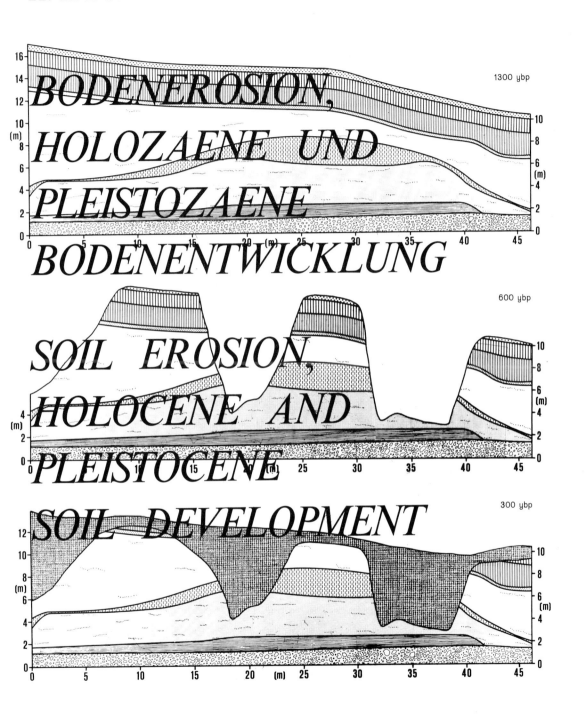

CATENA SUPPLEMENT 3

CATENA paperback

Joerg Richter

THE SOIL AS A REACTOR
Modelling Processes in the Soil

If we are to solve the pressing economic and ecological problems in agriculture, horticulture and forestry, and also with "waste" land and industrial emmissions, we must understand the processes that are going on in the soil. Ideally, we should be able to treat these processes quantitatively, using the same methods the civil engineer needs to get the optimum yield out of his plant. However, it seems very questionable, whether we would use our soils properly by trying to obtain the highest profit through maximum yield. It is vital to remember that soils are vulnerable or even destructible although or even because our western industrialized agriculture produces much more food on a smaller area than some ten years ago.

This book is primarily oriented on methodology. Starting with the phenomena of the different components of the soils, it describes their physical parameter functions and the mathematical models for transport and transformation processes in the soil. To treat the processes operationally, simple simulation models for practical applications are included in each chapter.

After dealing in the principal sections of each chapter with heat conduction and the soil regimes of material components like gases, water and ions, simple models of the behaviour of nutrients, herbicides and heavy metals in the soil are presented. These show how modelling may help to solve problems of environmental protection. In the concluding chapter, the problem of modelling salt transport in heterogeneous soils is discussed.

The book is intended for all scientists and students who are interested in applied soil science, especially in using soils effectively and carefully for growing plants: applied pedologists, land reclamation and improvement specialists, ecologists and environmentalists, agriculturalists, horticulturists, foresters, biologists (especially microbiologists), landscape planers and all kinds of geoscientists.

Prof.Dr. Joerg Richter
Institute of Soil Science
University of Hannover, FRG

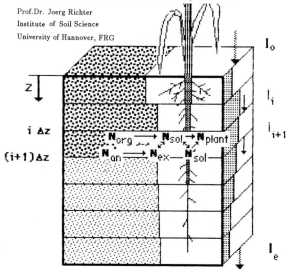

ISBN 3-923381-09-3   Price: DM 38,50 / US $ 24.—

FORTHCOMING PUBLICATION

publication date: June 1989

Heinrich Rohdenburg

LANDSCAPE ECOLOGY - GEOMORPHOLOGY

CATENA paperback

1989/about 220 pages/DM 44.- / US$ 28.-

ISBN 3-923381-15-8

ORDER FORM

Please send me .... copy/ies: Heinrich Rohdenburg, **LANDSCAPE ECOLOGY - GEOMORPHOLOGY**, CATENA paperback. 1989. DM 44.-/ US$ 28.-/ ISBN 3-923381-15-8

Name ....

Address ....

Date/ Signature ....

Please charge my credit card: ☐ Mastercard/Eurocard/Access ☐ Visa ☐ Diners

card no.: ....

expiration date: ....

signature: ....

CATENA VERLAG, Brockenblick 8, D-3302 Cremlingen-Destedt, West Germany, tel.05306-1530, fax 05306-1560

USA/CDN: **CATENA VERLAG**, P.O.Box 368, Lawrence, Kansas 66044-0368, USA, phone (913)843-1234 fax (913) 843-1274

SOIL TECHNOLOGY

A Cooperating Journal of CATENA

SOIL TECHNOLOGY

This quarterly journal is concerned with applied research and field applications on

- soil physics,
- soil mechanics,
- soil erosion and conservation,
- soil pollution,
- soil restoration.

The majority of the articles will be published in English but original contributions in French, German or Spanish, with extended summaries in English will occasionally be considered according to the basic principles of the publisher CATENA whose name not only represents the link between different disciplines of soil science but also symbolizes the connection between scientists and technologists of different nations, different thoughts and different languages.

The coordinator of SOIL TECHNOLOGY:

Donald Gabriels,
Faculty of Agricultural Sciences, State University of Gent,
Coupure links 653,
B-9000 Gent, Belgium (tel 32-91-236961).

Editorial Advisory Board:

J. Bouma, Wageningen, The Netherlands
W. Burke, Dublin, Ireland
S. El. Swaify, Hawaii, USA
K. H. Hartge, Hannover, F.R.G.
M. Kutilek, Praha, CSSR
G. Monnier, Montfavet, France
R. Morgan, Silsoe, UK
D. Nielsen, Davis, Californ., USA
I. Pla Sentis, Maracay, Venezuela
J. Rubio, Valencia, Spain
E. Skidmore, Manhattan, Kansas, USA

**Editorial Office
SOIL TECHNOLOGY**

Dr. D. Gabriels
Department of Soil Physics
Faculty of Agriculture
State University Gent
Coupure Links 653
B-9000 Gent
Belgium
tel. 32-91-236961

**Papers published in
Vol. 1, No. 1, March 1988**

S. A. El Swaify, A. Lo, R. Jay, L. Shinshiro, R. S. Yost: Achieving conservation-effectiveness in the tropics using legume intercrops.

I. Pla Sentis: Riego y desarollo de suelos afectados por sales en condiciones tropicales. / Irrigation and development of salt affected soils under tropical conditions.

K. H. Hartge: Erfassung des Verdichtungszustandes eines Bodens und seiner Veränderung mit der Zeit. / Techniques to evaluate the compaction of a soil and to follow its changes with time.

M. Kutilek, M. Krejča, R. Haverkamp, L. P. Rendon, J. Y. Parlange: On extrapolation of algebraic infiltration equations.

M. Šir, M. Kutilek, V. Kuráž, M. Krejča, F. Kubík: Field estimation of the soil hydraulic characteristics.

J. Albaladejo Montoro, R. Ortiz Silla, M. Martinez-Mena Garcia: Evaluation and mapping of erosion risks; an example from S. E. Spain.

SHORT COMMUNICATIONS

D. Gabriels: Use of organic waste materials for soil structurization and crop production; initial field experiment.

K. Reichardt: Aspects of soil physics in Brazil.

P. Bielek et al.: Internal nitrogen cycle processes and plant responses to the band application of nitrogen fertilizers.

V. Chour: An actual demand for improved soil technology in irrigation and drainage design in Czechoslovakia.

BOOK REVIEWS

ORDER FORM:

Please, send your orders to your usual supplier or to:

USA/CANADA: CATENA VERLAG
P.O.BOX 368
Lawrence, KS 66044
USA
phone (913) 843-12 34

Other countries: CATENA VERLAG
Brockenblick 8
D-3302 Cremlingen
West Germany
phone 0 53 06/15 30
fax 0 53 06/15 60

SOIL TECHNOLOGY 1988: Volume I (4 issues)

- ☐ please, enter a subscription 1988
  at US $ 120.— / DM 198,—
  incl. postage and handling
- ☐ please, send a free sample copy of **SOIL TECHNOLOGY**
- ☐ please, send guide for authors
- ☐ please, enter a personal subscription 1988 at 50 % reduction
  (available from the publisher only)
- ☐ I enclose ☐ check ☐ bank draft ☐ unesco coupons
- ☐ charge my credit card (only for orders USA/CANADA)
     ☐ Master Card   ☐ Visa

Card No. _____
Expir. Date _____
Signature _____
☐ please, send invoice
Name _____
Address _____
Date/Signature _____